THE COLORADO

RIVERS OF AMERICA

THE COLORADO

FRANK WATERS

Illustrated by Nicolai Fechin
Maps by George Annand
With a new foreword by Jonathan Waterman

Guilford, Connecticut

An imprint of The Rowman & Littlefield Publishing Group, Inc.
4501 Forbes Blvd., Ste. 200
Lanham, MD 20706
www.rowman.com

Distributed by NATIONAL BOOK NETWORK

Copyright © 2021 The Frank Waters Foundation
Foreword copyright © Jonathan Waterman

Illustrated by Nicolai Fechin
Maps by George Annand

All rights reserved. No part of this book may be reproduced in any form or by any electronic or mechanical means, including information storage and retrieval systems, without written permission from the publisher, except by a reviewer who may quote passages in a review.

British Library Cataloguing in Publication Information available

Library of Congress Control Number: 2020942146

ISBN 978-1-4930-4014-8 (paperback: alk. paper)
ISBN 978-1-4930-4015-5 (electronic)

∞™ The paper used in this publication meets the minimum requirements of American National Standard for Information Sciences—Permanence of Paper for Printed Library Materials, ANSI/NISO Z39.48-1992.

Contents

Foreword	vii
Preface to the 1984 Edition	xvii
Introduction: Its Character	xxvii

Part One: Its Background

1. The Colorado Pyramid	3
2. High Country	17
3. The Shining Mountains	31
4. Mesa and Plateau	53
5. The Desert	81
6. Its Delta	101

Part Two: Its People

1. The Conquerors	131
2. The Padres	145
3. The Trappers	164
4. The Settlers	185
5. The Outcasts	203
6. Its Travelers	224
7. "The People"	246
8. The Inheritors	269

CONTENTS

PART THREE: ITS FUTURE

    1. IMPERIAL VALLEY 287
    2. THE COLORADO RIVER PROJECT . . 317
    3. BOULDER DAM 329
    4. CANAL, AQUEDUCT AND TREATY . . 344
    5. LONG VIEW 359

PART FOUR: GRAND CAÑON

REFERENCE APPENDIX 381

GLOSSARY 387

Index 389

MAPS

THE COLORADO . . . . FRONTISPEACE
HIGH COUNTRY 16
THE MOUNTAINS 32
MESA AND PLATEAU 54
DESERT COUNTRY 80
DELTA COUNTRY 102

Foreword
by Jonathan Waterman

In these plethoric times, American rivers are often bypassed by asphalt highways, harnessed by levees and dams, or overlooked for their natural beauty. Yet as a primary source of drinking water, wild waterways also support the entire web of life. To those who live alongside, visit, or travel on rivers, the esthetic and spiritual values are priceless. And these rivers literally carry our cultural history — think Washington crossing the Delaware, Lewis and Clark exploring the Missouri, Huck Finn lighting out on the Mississippi, or John Wesley Powell mapping the Colorado.

More than eighty years ago, to celebrate these American thoroughfares — built of two hydrogen atoms linked to one oxygen atom, and multiplied infinitesimally — a New York publishing house, Farrar & Rinehart, went to work on an immense and enduring project. Meanwhile, the Bureau of Reclamation and the Army Corps of Engineers were already busier than beavers on said rivers.

In 1937 Farrar & Rinehart released *Kennebec: Cradle of Americans*. Its author, Robert P. Tristram Coffin, had already written a dozen books of prose and nonfiction. Showing the artistic métier for the river authors who would follow, Coffin had just won the Pulitzer Prize for poetry.

Kennebec was the first book amid a veritable literary river system, conceived by the author and theater critic Constance Lindsay Skinner. As Farrar & Rinehart's editor of the series, she aimed to interpret America through the history, exploration, and flow of its rivers. Rather than appointing river experts, she sought out prose stylists and scholarly novelists with historical sensibilities to chronicle the big picture of regions surrounding America's rivers. She wrote, "This is to be a literary and not a historical series." The plan, she said, was to capture "the folk life."

The literary talent would include such luminaries as Marjory Stoneman Douglas. Her poetic reportage in *The River of Grass* would help preserve the Everglades. Then Paul Horgan's *Great River* about the Rio Grande won the Pulitzer Prize for history. The series would also include renowned illustrators—the budding young artist Andrew Wyeth illustrated number 13 in the series *The Brandywine* (1941, written by the critic, editor, and Yale University professor Henry Seidel Canby).

While at her desk putting the final touches on *The Hudson*, Skinner collapsed and died from a heart attack. Undiminished, *The Hudson*—written by the best-selling author Carl Carmer, who would became one of four editors for the river books—numbered sixth in a series that the ambitious Constance Lindsay Skinner had envisioned to include a total of twenty-four books. But as the series became popular, it kept, shall we say, *flowing*. Ultimately, it included sixty-five iconic American rivers. All were published under the Rinehart name until 1974.

Eventually, the series became collector's items, spawning a dozen other "Series Americana" (including *American Trails, American Folkways, Regions of America*, and a little-known young adult series of river books). In 1997, the surviving authors and illustrators of the river books were celebrated at the Library of Congress during the sixtieth anniversary of the series. In 2001, the two volume, 901-page bibliography of the American Rivers Series was published.

Today, as this nation of rivers ponders the wisdom of trying to control waterways subjected to increasing floods or drought, Lyons Press (under the aegis of the editor Rick Rinehart, grandson of the original publisher) has reprinted *The Missouri, The Connecticut*, and *The Colorado*. As the thirtieth book in the original series, *The Colorado* was first published in 1946 by Rinehart & Company. Its author, Frank Waters, from New Mexico, spent five years researching and writing the book while he worked for the Office of Inter-American

Affairs as a propaganda analyst in Washington, DC, during World War II.

Waters, repeatedly nominated for the Nobel Prize in Literature for his novels about Native American life in the American Southwest, had recently found a captivated audience with his fictionalized account of a young Pueblo hunter arrested for shooting a deer in the New Mexico National Forest two days after hunting season had ended. This book, *The Man Who Killed the Deer* (1942), helped persuade the government to return 48,000 acres of sacred Puebloan land to the tribe. In the sixties his *Book of the Hopi* developed a cult following among spirituality seekers. Yet Waters—ahead of his time as an environmentalist and a mystic—would remain one of America's best known unknown writers in a time when Native American literature remained an obscure topic.

Until *The Colorado*, there were no books of reportage about the river. (John Wesley Powell's *Exploration of the Colorado River and Its Canyons* was published in 1895 but was more an adventure narrative about a then unknown part of the continent.) Today, of course, there are scores of books reporting on the environmental and water-shortage crises of a river often referred to as The American Nile.

From its source in the Colorado Rocky Mountains, the Lower Forty-Eight's most precipitous river drains 242,000 square miles and plummets 14,000 feet toward the vast trunk of river delta in what Waters refers to as the Colorado Pyramid. Despite more than a hundred dams now cemented throughout the Colorado River Basin (Waters wrote about the nine dams that existed in the 1940s), the flooding river still defies its confinement. In the mid-twentieth century, several species of Colorado River fish found nowhere else in the world were not yet endangered, the river still ran to the sea through its delta, and there was scant industry and sparse population. Waters reported 616,000 people living in the state of Arizona; today its population is over 7 million.

Even during World War II, the culture and lives of two dozen different Native American tribes who lived along the river were already fractured. This includes Kwapa, Cocopah, Quechan, Mohave, Chemeheuvi, Ak-Chin, Ouray, Zuni, Kaibab, Papagos, Pueblo, Pima, Apache, Hualapai, Havasupai, Navajo, Uintah, Ute, Hopi, and Paiute. In this respect, Waters, part Cheyenne, may have been the most qualified man anywhere to write about the river.

"We were both born in the high Colorado Rockies," he wrote. "Progressively in childhood and youth we made our way down the peaks and mountains. Meandering back and forth across mesa and plateau our lives assumed their permanent color, our tempers set. On the desert below we both were harnessed to work for the first time. And the last part of this vast background that I saw, like the river, was its Mexican delta."

Until his dying day in 1995, Waters believed that a conventional Western European outlook could not understand the environment surrounding the Colorado River. Waters asserted that it took a patient, intuitive person respectful of the land—in other words, a Native American.

In his preface to a more recent edition, Waters wrote: "We are discovering that ecology always has been the core of Indian religious belief.... Our Mother Earth, fructified by Father Sun, gives birth to all her children, which are given by air the breath of life, and are nourished by the waters." Today the *Mother Earth* vernacular might be perceived as a cliché, but more than seventy years ago Waters—who had become a New Ager before the term even existed—had captured one of many paradigms for those who live close to nature.

While many have tried in vain to capture the river's essence, the appropriately named Waters refers to it simply as an outlaw. "As no other, it is savage and unpredictable of mood, peculiarly American in character. It has for its background the haunting sweep of illimitable horizons, the immensities of unbroken wilderness."

In *The Colorado*, Waters breaks the region and its cultural history into four parts, beginning with "Its Background," dropping from the apex of the pyramid though the high country, the shining mountains, mesa and plateau, desert then its delta. To the original indigenes of the west, the shape of the world was essential, Waters wrote, since pyramids are a universal symbol of strength and durability—such as the pyramid on the Great Seal of the United States (used to authenticate federal documents). "The Colorado Pyramid," he wrote, "carved out of the continent, might well be the mother shape, the ancient and eternal prototype."

As his narrative descends through the plateaus, he described ancestral Puebloan ruins "that stand as if in a vacuum; like little pink stone castles in a child's glass ball." Into the desert, which he describes as the antithesis of the mountains, he related his experiences working as a lineman in what felt like "molten" heat. Throughout, his prose descriptions of the landscape are interspersed with local vernacular about farting mules and crusty characters still found about the river basin today.

Such as an old desert rat describing the land: "Pert near anybody can tell you what it ain't," as the author captured the quote, "there ain't nobody can tell what it is." Yet Waters, a masterful storyteller, shows us exactly what the river *is*.

Particularly in Chapter Six, "Its Delta," a highlight of the book. Describing the base of the Colorado Pyramid as 2,000 square miles of "wild terra incognita" with abundant game amid "a crazy quilt of waterways," he tells the forgotten story of the once mighty bore tide that swept up from the Gulf of California and into the river delta. In November 1922 a 36-ton steamer made its way north up the river from the sea with 125 farmhands bound for Mexicali. Cabled to the shore for the night with its passengers sleeping on board along the now abandoned hamlet of La Bomba, everyone on board was "awakened by a terrifying roar. It sounded like a gigantic waterfall booming downriver." Those who rushed on deck under the moonlight saw a fifteen-foot-high wall of water, and with "hardly time for a cry" the tidal bore snapped the two steel hawsers anchoring the big boat to shore and rolled them all over. Eighty-six passengers drowned.

And three years later, after a Mexican friend suggested that he take the trip for a "pleasant" stroll, Waters himself boarded an "unpainted, repellent and inexpressively dirty" steamer called the *Rio Colorado*. His adventure riding out the bore tide, along with his description of his fellow passengers and riverscape that has long since been lost, remain a classic of Colorado River literature.

* * *

More than eighty years later, living in the headwaters of the Colorado River, I too had a desire to experience the delta. So starting in June of 2008 I took a journey, often alone, from the headwaters in northern Colorado to Mexico. Along the way I learned that the river touches many lives. As Frank Waters believed, the story of the river has to be told through the people who live with it. Among this culture, I met a belligerent rancher who tried to kick me off the river where it flowed past his land. I interviewed engineers devoted to reclaiming the Colorado River water. Or boatmen in the Grand Canyon who held forth on birds, geology, and history—all while navigating rapids. I took notes while listening to a Las Vegas water manager spout out acre-feet calculations as quickly as a blackjack dealer slings cards. I met a farmer in Imperial Valley saving river water by installing drip irrigation. And repeatedly, I saw how the drying river had a devastating ripple effect among Native American communities. Near the end of my journey, five months out, I experienced the Colorado River's greatest tragedy, one that Frank Waters fortunately didn't live to see. In January at the Mexican border I paddled past plywood shacks and children flipping stones into the narrowing stream, holding my breath, hoping the river wouldn't stop. Upstream, on the border bridge, trucks free of mufflers tore through the morning haze, redolent of burning garbage. Two miles into Mexico, my hopes of a complete 1,450-mile descent ended in a foamy pond of congealed fertilizers, distillate of countless American lawns and 3.4 million thirsty farm acres. I splashed out in bare feet, worried that our most iconic white-water river would make me physically ill and tried to wipe the river muck off my pack raft with tamarisk fronds, cursing the system that has diminished the mighty Colorado to a stinking cesspool. Then I deflated and folded the little boat onto my pack and began walking the route Waters once rode in a steamship. Bushwhacking and trudging through sands washed from the Rockies, baked by the hot sun, I perspired faster than I could drink as I stumbled a dozen miles south into Sonora, Mexico. The sun faded over the distant Sierra del Mayor, dappled an incandescent-carrot hue through steaming Mexicali air pollution, and I collapsed a dozen feet above one of many dried-out stream banks. The Colorado River, I had learned, has been engineered to death. Its plethora of dams and canals divert its water to most every farm, industry, and city within a 250-mile radius of the river. Each year, seven western states and northern Mexico take 16.5 million acre-feet (enough water to supply 33 million American households) of river water. Amid a mega drought in the southwest, climate models show that conditions will continue to dry the snowmelt-fed river. Add explosive population growth, increasing the demand for water, and the river's future becomes a ticking time bomb.

This was my sleepless perspective from the 3,000-square-mile Colorado River Delta, being subsumed by the Sonoran Desert. Distant dogs howled with hunger, while the northern horizon burned white-hot with the international border's halogen lights. Since 1998, drought and overuse of the river have stopped it from flowing across this border to the sea. For most of the final 70 miles to come, I would be walking.

Over the next few days, I guzzled over 50 pounds of water, shifting the weight from my back to belly and stung by the knowledge that a river was supposed to flow where I staggered through brush and poured sweat and drank water imported from far-off aquifers. Occasionally the river reemerged in stagnant ponds shaded by cottonwoods and guarded by reluctant great blue herons, icons of a former cornucopia.

I wandered for ten days, southwest into Baja California, then south toward the Sea of Cortez. Most of the time I was half lost in the dried-out maze of delta cut by farm fields, salty canals, potholed tarmac, and railroad tracks. Eventually, a small tributary, El Rio Hardy, acted as delta resuscitation. Meanwhile, a newly developed waterborne infection in my blistered and red swollen feet had me hobbling.

So I paddled the Rio Hardy before it could be sucked under the vast delta. On the second day afloat, gobsmacked, I found the wet paradise that Frank Waters had described in his book. The glowing, green-phosphate water turned clear, scrubbed clean by a rowdy coiffure of reeds and plants.

These curlicues of hidden river were lush with an upwelling of underground water, temporarily arisen before it would be reabsorbed and blocked from the ocean by ancient sand grains—spread as far as I could see—carved from and carried 600 miles out of the Grand Canyon. Here, briefly, nature endured: rattling kingfishers, squadrons of circling mallards, and hushed, stern-faced cattle egrets. I could smell the postcoital tang of ocean tides.

Tamarisk thinned. Salt grass bearded the ground. Pintail ducks, curlews, ibis, plovers, and black-crowned night herons fluttered and gabbled and splashed. Sere mountains surrounded me, below an infinite sky, bisected by a once unstoppable river that knew no banks. As the stream narrowed, I could feel it gathering momentum, as if it would once more meet the sea.

But I had to stop holding my breath. I had developed a fever from the infection in my feet. And I would never find the lush delta that Waters had so aptly described.

* * *

After Water's luminescent musings on the delta—since the American River Series was conceived as a literature that would transcend history with a sense of place and culture—Part Two of *The Colorado* delves into "Its People." From the Spanish conqueror Ulloa (who first recorded a massive 36-foot tide but turned back short of the delta and its walls of water) probing up the Gulf of California, Waters details the amazing story of the Black slave Estevan, who wandered up the river into sixteenth-century Arizona, into the lands of the Navajo and Apaches, until the Zuni people cut him into small pieces. Flowing through subsequent chapters on "the Padres," "the Trappers," and "the Settlers," Waters continues his storytelling prowess, painting a thorough picture of the people and the great red river of the west. In the "Outcasts," the author gives broad brushstrokes of the gold rush desperados and then takes the reader on a fascinating tangent with Wyatt Earp—decidedly not a hero in Water's view—abandoning Arizona and various wives en route to California. In "Its Travelers" he relates more forgotten nineteenth-century yarns about Lieutenant Hardy (who named the only Colorado River tributary remaining in the delta) and wrecked steamboats. Then he delves into those he considers the real "People," which is the derivation of most Native American tribal names—Navajo, Cheyenne, Apaches, or Choctaw. Finally, and wistfully, in "The Inheritors," Waters describes modern times and what we (the author does not exclude himself from this) have lost in our separation from the environment. Part Three, "Its Future," allows Waters once again to share his time in the Imperial Valley and show the emergence of agriculture in the river basin. This is then juxtaposed in ensuing chapters with the actual engineering, damming, and canaling of the river. To be an environmentalist at the close of the greatest war on earth, when America was steadfastly subduing the natural world, defined Waters as both an outcast and a visionary. "For all our technological achievements," he wrote, "our very lives tremble upon the delicate scales of nature." His decision to end the book with a single Part Four chapter, "Grand Cañon," shows Waters ending on a note of hope. After all, there is nowhere else in the Colorado Pyramid where a traveler, or Waters's readers, can glimpse such a massive and enduring chasm of still pristine, ancient earth. As always, his words describe it best: "Grand Cañon is the world's largest and oldest book.... Though its pages are wrinkled, creased and worn, they are brilliantly colored and beautifully engraved. A few chapters are missing. But so clearly are the others written that their meaning is revealed without break in continuity." In the next single-sentence paragraph, he implores, "Thumb down through its rock pages." Which should be taken as an enticement to read Waters's own timeless work.

PREFACE to the 1984 Edition

This preface to the second reissue of *The Colorado* is substantially the same as that written for the 1974 edition. Its minor revisions do not presume to bring it up to date, and it still seems both necessary and unnecessary.

Since the book itself was published almost forty years ago, the face of the 246,000-square-mile drainage area of the Colorado River, a heartbreakingly beautiful wilderness, has changed beyond recognition. It is no longer the "most sparsely settled area of its size in the Western Hemisphere." Phoenix, Arizona, is not now "its sole city, the only community of 50,000 population and not yet a city by metropolitan standards." Greater Phoenix is a metropolis exceeding 1,110,000. Tucson has more than 337,000 permanent residents. One-street, one-hotel Yuma—"Free meals every day the sun doesn't shine!"—is now a modern city with air-conditioned hotels and motels outspread for miles over the desert.

Postcard Arizona is outmoding the statistical facts about it faster than they are reported. In 1946, when this book was first published, the state's population was 616,000. By 1980 it had grown to over 2,717,000. Something else has happened with this great influx of people. Arizona is no longer developing solely through its former three economic "C's": cattle, cotton and copper. It is emphasizing the growth of industry. In 1946 its manufacturing output was $90 million. By 1980 this had risen to $3.9 billion, exceeding the production of agricultural crops, livestock, and mining.

What has happened in Arizona is matched by changes in the portions of the other six states in the Colorado drainage area: Wyoming, Colorado, Utah, New Mexico, Nevada, and California. Las Vegas, Nevada, "Capital of the Land of the Hard-Way Eight," has emerged with its great Strip pleasure palaces as the major center of open gambling and frenetic entertainment in the United States. Aspen, Colorado, is only one of a dozen old Rocky Mountain mining "ghost towns" that have come to life as fashionable ski resorts. And so it is everywhere: small towns grown into cities, new towns, tourist resorts, and industrial centers springing up beyond count.

In former days we drove through the country over rutted dirt roads, generally unmarked and lined with blown-out tires. On the running boards of our cars were mounted tin containers divided into three painted compartments: red for gasoline, blue for oil, and white for water. Standard equipment included a couple of blankets, a coffee pot and frying pan, and emergency rations. They were usually needed. There were no bridges over the countless "dry" arroyos that would be filled with roaring torrents from flash floods. One was compelled to camp until the water died down. Traveling had great advantages also. We could pull off the road anywhere of an evening to camp, our fire drawing travelers for company.

These crawling dirt roads have now been replaced by a network of high-speed highways with underpasses, overpasses, and cloverleafs bounded by hurricane wire fences. There is no place to turn off for a picnic. One has to hurtle on to a neon-lit hamburger and hot-dog emporium miles ahead.

The great Santa Fe Railroad that spanned the Southwest by crack fliers with wonderful names—the Chief, Navajo, El Capitan, and Grand Canyon Limited—has been outmoded by cars and airplanes. The incomparable Fred Harvey Houses, spaced mealtime-distance apart along its route to provide food and lodging, are but a nostalgic memory. Huge airports have replaced their function, but not their charm and architectural relationship to their landscape.

The little-known ruins of prehistoric America are overcrowded with visitors. The location of majestic Pueblo Bonito in Chaco Canyon, New Mexico, built about a thousand years ago, was tagged as "one hundred miles from anywhere." Today, by paved roads, it is easily accessible to motorists who come for a quick look. Tens of thousands of visitors are wearing down the trails to the fabulous cliff dwellings at Mesa Verde, Colorado, and sonic booms are weakening their walls.

PREFACE   xix

Indian reservations are no longer remote little countries inhabited by strange peoples speaking their own tongues, wearing their distinctive costumes, observing their customs and traditions. The isolated trading posts, once their only contact with the world outside, have vanished.

The immense Navajo Reservation of 25,000 square miles now boasts nearly 160,000 people with a tribal government of their own and a modern capital at Window Rock. With the development of its natural resources—timber, oil, coal, and minerals—the tribe is the largest and richest in the nation. It publishes its own newspaper and broadcasts its own radio programs in Navajo.

The Navajo Nation's greatest boost to affluence came after this book was written. On July 16, 1945, the first experimental atomic bomb was detonated in the desert of southern New Mexico, ushering in the Atomic Age. The only available sources of uranium at the time were in Brazil and Canada. Shortly thereafter Paddy Martinez, a Navajo, brought into town a piece of red rock from Haystack Butte, a few miles north of Grants, New Mexico. Tests determined it to be carnotite, a uranium ore. The rush began—a rush comparable to the great gold rushes of the century before. Explorations revealed that the whole Colorado Plateau contained one of the most extensive uranium deposits in the world. Wilderness areas were opened up for mining. Cowtown Grants, a block long, lengthened into miles as it ballooned into the uranium capital, with other industrial complexes in Colorado, Utah, and the nuclear test sites in Nevada, "The Land Where the Giant Mushrooms Grow."

Industrialization spread with the discovery of great beds of bituminous coal in the Four Corners region—the only place in the United States where four states converge: Utah, Colorado, New Mexico, and Arizona. Strip-mining on Black Mesa, sacred to both Hopis and Navajos, was begun by a consortium of the nation's largest power companies with the approval of the federal government. The Four Corners plant was the first of six coal-burning power plants to be erected in the area. Its thirty-story smokestacks spewed out more pollutants than Los Angeles and New York combined, covering thousands of square miles.

The strip-mined coal was pulverized, mixed with water, and pumped through a slurry pipeline to the Mojave Plant in southern Nevada. The necessary water was drawn from underground at the rate of 2,300 gallons a minute, lowering the water table of this arid desert and drying up the springs and waterholes upon which the Indians were dependent. The permanent destruction of the land by strip-mining, pollution of the air, and exhaustion of the water, with miles of transmission lines, work roads, railroad spurs, and pipelines—the "Rape of Black Mesa" by greedy private interests—roused a storm of public protest. Environmental legislation was instituted to reduce the pollution and damage. Yet still more immense power plants were erected.

As for the Colorado itself—that wild and turbulent river which threaded this once incomparable wilderness from 14,000-foot-high peaks to a desert 248 feet below sea level—it is no longer a river of mystic beauty and sublime terror. It is virtually a cement-lined irrigation ditch from source to mouth. Of its average annual flow of 13 million acre-feet, an acre-foot being equivalent to the amount of water necessary to cover one acre with one foot of water, not one drop now flows from its mouth into the Gulf of California.

Boulder Dam, completed in 1935, was the first major dam on the Colorado, and the largest in the world yet built. A mammoth technological marvel, it evoked my own extravagant praise and admiration with that of the entire country. Since then I have come to regard it as the first of our misguided attempts to dominate the entire natural world of the river.

There are now nine major dams which control the Colorado's flow, with backed-up lakes to serve as storage reservoirs, hydroturbine plants to generate power, aqueduct systems, and pumping works. In addition to these mainstream dams are others located on the Colorado's major tributaries. Their costs are borne largely by the lucrative sale of electric energy to the ready markets of Las Vegas, Phoenix-Tucson, and the megalopolises of southern California. As this became known, it aroused uneasy suspicions that the building of still more dams and power plants would generate still more power for which a profitable market could be created by developing demands for electric toothbrushes, backscrubbers, and other unnecessary gadgets.

PREFACE

These speculations were given public voice when still another federal dam in Marble Canyon was proposed. With the Colorado already controlled by nine major dams, why should a tenth be constructed unless for the lucrative sale of power from its hydroelectric plant? This consideration was overshadowed by the fact that the Marble Canyon Dam would flood one of the greatest natural wonders of the world—the incomparable Grand Canyon of Colorado itself. Sparked by the Sierra Club of California, nationwide public protest effectually stopped this catastrophic sellout, at least for the time.

Industrialization of all the Colorado basin is now included in the avowed federal program to economically exploit all the natural resources of the entire nation under the press of an "energy crisis." What the disastrous results of this short-sighted policy will be, no one knows. But certainly more drastic changes are under way. To detail them here would necessitate rewriting this entire book—not only this year, but every year hereafter to keep it up to date.

In greater perspective the recording of these changes is unnecessary. *The Colorado* was not written as a cursory guidebook to points of interest in a little-known land. Nor was it imagined that it would ever be read as a nostalgic reminder of the past. If I can condense its intent into a few words, the book was an attempt to perceive, from the viewpoint of the eternal rather than from that of time, the presence of the spirit-of-place of the immense wilderness of the Colorado and its effect upon us.

How valid is this viewpoint today?

Surely each of the continental masses of the world has its own great spirit-of-place, imbuing it with a distinctive and ineradicable rhythm, mood, and character. Every country, too, reflects the psychic influences of its own earth. It has been reported that the skull measurements of all European races immigrating to America begin to change in the second generation. C. G. Jung, the Swiss psychologist, discovered so many Indian symbols in his analyses of patients that he concluded the American was a European with an Indian soul. A mysterious physiological and psychological adaptation of man to his environment.

This vast heartland illustrates it well. Many races and many breeds of men have reflected its influence in their religion and secular organizations, their very lives: the trappers and mountain men, explorers, outlaws, prospectors, river-running adventurers, and settlers. All in their ways reacted to its dictates.

Are we today immune?

In our march of conquest westward from the Atlantic, we Anglo-Americans viewed the earth as an inanimate treasure house existing solely to be exploited for our material gain. With a calculated program of genocide we exterminated almost all Indian tribes and wantonly wiped out all animals and birds save a few remaining species. We leveled whole forests under the axe, plowed under the grasslands, dammed and drained the rivers, gutted the mountains for gold and silver, divided and sold and resold the land itself. But that which was destroyed and stolen from nature was at the same time paid for by loss of soul, by our alienation not only from the earth, but from our dark maternal unconscious, its psychic counterpart.

The high barrier of the Rockies long retarded the advance of industrialization. Even when this book was written, the Colorado heartland remained a virtual wilderness in which its sparse population was necessarily attuned to the land and the forces of nature in contrast to the dense population of the East, which was more oriented to social forces. Then, as the flood poured in to this last oasis, we began to be aware of the results of our savage onslaught against nature.

Country-wide the topsoil had been denuded. Lakes and rivers had been polluted, and the underground water level was lowering. Even the seas were being contaminated by wastes and refuse. The very air we breathed was becoming so dangerously toxic that programs were being set up to evacuate large cities on notice, and radioactive fallout was laying wide swaths around the entire planet. We had become the richest and most materialistic nation in history, yet our civilization was on the verge of collapse.

Hence our belated recognition of ecology, defined by Webster as "biology dealing with the mutual relations between organisms and their environment." Strictly in accord with this, we are beginning to take steps to halt further pollution of land, water, and air by resistant industries and greedy private interests. Strangely coincidental with our interest in ecology has been our awakened interest in Indians—and especially those in the Southwest, which are virtually the only tribes left as integral groups within their immemorial homelands.

Perhaps this coincidental interest is not so strange after all. We are discovering that ecology always has been the core of Indian religious belief. But how different it is from our understanding of ecology from a solely physical or biological level: "the mutual relations between organisms and their environment." We have not yet comprehended, as have the Indians, the psychical ecology underlying the physical. The earth is not inanimate. It is a living entity, the mother of all life, our Mother Earth. All her children are alive—the living stones, the great breathing mountains, plants and trees, as well as birds and animals and man. All are united in one harmonious whole. Whatever happens to one affects the others and subtly changes the pattern of the whole. For all these living entities, like man, possess not only an outer physical form, but an inner spiritual component. An Indian may fell a pine or kill a deer in order to utilize its physical form for his material needs. But aware of the need for invoking its spiritual life as a source of psychic energy, he must first request ritual permission for its sacrifice. This custom has been followed since ancient times in all Indian America.

Indian ceremonialism is founded upon reverence for the primary elements of life—earth, air, water, and the creative fire of the sun. Our Mother Earth, fructified by Father Sun, gives birth to all her children, which are given by air the breath of life, and are nourished by the waters. . . . It is not as simple and primitive as we suppose.

Theodor Schwenck in Switzerland, in his illuminating study of water, reminds us that water is the main constituent of the body of man as it is of the body of this planet. He believes, like the Indians, it is a living element with a spiritual nature, in tune with the phases of the moon, sensitive to all planetary influences, and thus relating us to the greater body of the living universe. Rivers spread out at the full moon and narrow at the new moon; hence lumbermen float their logs downstream at the new moon lest they be beached on the banks during the full moon. Sap in trees varies also; timber is felled at the new moon in winter. Schwenck's revealing photographs show that running water always tends to take a spherical form, a totality, in its meandering loops and spirals. And this is reflected by all living creatures in the archetypal forms of their bones and muscles, in the spiraling shapes of antelope horns, in man's own bloodstream.

So it is with air. It ascends over warm land, forming low-pressure areas, and descends over cool land, forming high-pressure areas. These alternate with day and night, with the seasons, constituting the "breathing of the continents."

And the sun. Man breaths 18 times a minute, 25,920 times a day. Hence he is connected with the sun, for 25,920 years is the duration of the great cycle of the Precession—the time it takes for the vernal equinox to move through the circle of the zodiac. For every 18 breaths there are 72 beats of the pulse, a ratio of 1:4 for the circulation of blood. As the propagation of sound is four times faster in sea water (akin to blood) than air, our blood is the archetypal organ of liquid flow in man, akin to that of water throughout the earth.

So it is that by obstructing the free action of these elemental forces and ruthlessly destroying nature, man, who is also a part of nature, ruptures his own inner self. He alienates his conscious self from the earthly substratum of his essential being, the unconscious. Jung views this as the tragedy of overcivilized man. It is the disease of our time.

The impending tragedy was of course not apparent forty years ago in this last great wilderness of America, its spiritual heartland. Then one could experience the palpable presence of its mysterious spirit in its mighty peaks, its waves of forested ranges, the isolate salmon-tinted buttes and mesas, in its sublime canyons, and turbulent rivers. Here one knew the last Indian tribes remaining in their ancient homelands, acknowledging it in their mystical ceremonies whose ritual dances, songs, and recitals constitute America's only indigenous Mystery Plays. It was this feeling, this perception of the eternal verities of the land, that impelled the writing of this book.

So precious is this spiritual heartland, it seems to me now, that it would not have been inappropriate had we with better foresight preserved it as a refreshing oasis, a national shrine, for the millions of people desperately needing to regain touch with their earth and their inner selves. Who can say that its spiritual and aesthetic values would not in the long future have outweighed its monetary values?

Nebulous as this psychic factor in the human equation may seem to us, it should be included in programs now being projected to further increase the technological destruction of the land to meet the needs of our increasing population.

An ancient Inca proverb puts it neatly: "The frog does not drink up the pond in which he lives."

Despite all that has happened, the pond has not yet been completely swallowed. The subtle and indestructible spirit-of-place still hovers over us as it always has and always will. These pages hopefully suggest we ask ouselves whether it or we will write the epilogue to the long future.

FRANK WATERS

Introduction: Its Character

MOST rivers are confined to the needs and histories of men. Like roads, they seem inconsequential without their travelers. The Colorado is an outlaw. It belongs only to the ancient, eternal earth. As no other, it is savage and unpredictable of mood, peculiarly American in character. It has for its background the haunting sweep of illimitable horizons, the immensities of unbroken wilderness. From perpetually snow-capped peaks to stifling deserts below sea level, it cuts the deepest and truest cross section through the continent.

As the Rocky Mountains are the backbone of physical North America, the Colorado is the vertebral tube carrying the spinal fluid of the continent. From this viscous, reddish flow the river derives its name. Despite a score of other names, it has become known at last simply by its one unchanging color—in Spanish the *Rio Colorado,* the great Red River of the West.

Its landscapes are never anywhere urban or commercial, not even pastoral. They are purely mystical in tone. There are the wind-swept rocky wastes high above timberline, the sunless gloom of deep gorges. When the river does rise to the surface again it is upon the face of an earth whose expressions are never twice the same.

The black volcanic picachos [1] creep closer in the moonlight, baring their saw-tooth fangs. By day the crinkled desert hills diminish and recede, or merely float, bottomless,

---

[1] Unfamiliar Spanish and Indian words are defined in Glossary, p. 395.

upon the horizon. More often than not the mountains are mirages. Glistening salt beds and alkali flats turn into seas; uncovered veins of legendary native gold into mere banks of micaceous gravel.

In this shifting realm of the fantastic unreal only the river is permanent. It is the one enduring mesmer from whose spectral spell no man who has once seen it is ever quite freed.

Those who love it best are those who fear it most. For like all things touched with the sublime it carries a lurking horror, and its mysteries wear the mask of the commonplace. To allude to it as something more than a river would sound like a literary affectation only to the literary. The illiterate might well comprehend most fully all that it expresses. As from a Navajo sand painting or ceremonial blanket he would read the river's cryptic meaning in the earth it threads.

The dreamlike vacuousness, wild beauty and barbaric boldness of design form but the pattern of the warp which underlies the subtle, inimical resistance of the woof. Old, ancient America! With its own great spirit of place; with the shadows of aboriginal ghosts still gliding across it; and with its own demons not yet appeased—the haunting promise of the far-off, its tormenting unrest.

It is still a wilderness. To understand it you must think in new dimensions. You must feel in terms of depth as well as space, of eternity and not of time.

The
COLORADO

PART ONE

Its Background

CHAPTER I

The Colorado Pyramid

THE Colorado River drains an area greater than that of all Spain—which sent the first explorers into the region—and Portugal combined; and Belgium could be tucked into its great cañons for good measure.

This immense drainage area of 246,000 square miles covers nearly one-twelfth of the United States and a part of Mexico. It includes portions of Wyoming, Colorado, Utah and New Mexico in its upper basin; and parts of Arizona, Nevada and California in its lower basin. In its vast delta are included two portions of Mexico: a part of Sonora, on the mainland; and a part of the peninsula of Baja California, or Lower California, as distinguished from Southern California, which is part of the United States' mainland state of California.

It is the most sparsely settled area of its size in the Western Hemisphere.

All the familiar cities of the American West lie outside its vast wilderness: Cheyenne, Wyoming, beyond it to the north; El Paso, Texas, to the south; Salt Lake City, Utah, to the west; Denver, Colorado, and Albuquerque and Santa Fe, New Mexico, to the east.

Phoenix, Arizona, is its sole city—the only community

of 50,000 population in the entire quarter-million square-mile basin, and it is not yet a city by metropolitan standards.

In the 180,000 square miles above Parker there are but 115 incorporated towns and villages, one to each million acres. Just one of these has a population of 10,000; only five exceed 5,000. A total population of a mere 315,000 —an average of 1.7 per square mile compared with 45 for the United States as a whole.

Such a gaping vacuum is incomprehensible to a traveler from Europe. Paraguay with the vast pampas of the Gran Chaco, the country with the lowest density of population in the New World, 6.9, has over twice the population though only three-fifths as large. Bolivia on the roof of the Andes, ranking next from the bottom with 8.5 density, is only five-thirds as large but has seven times the population. Even to a commuter between Washington and New York it is difficult to realize, for in his afternoon ride he can look up from his newspaper every half hour and see outside seven successive cities each holding a population nearly equal to or much greater than that of the whole Colorado River basin.

This is a country where the land predominates, not man. It contains the highest peaks, the largest mountain ranges, the widest plateaus, the deepest cañons and the lowest deserts in America. Geology here forever dominates life and gives it its ultimate meaning. And this meaning is as pertinent to us today as it has been in the past, and will be in the centuries to come.

It is something of a paradox of terminology that this vast basin of the Colorado is also one of the greatest mountain masses in the world. In shape it is not a basin at all. It is a great, irregular stone pyramid. One side, its eastern wall, rears abruptly and almost a sheer two miles in height from the flat plains. From the deserts to the south and west it rises in great recessed terraces like the steps of a Toltec pyramid. Only to the north does it fall away gradually to the Arctic Circle.

Its granitic composition is nakedly exposed in the summits of the fifty and more peaks protruding above 14,000 feet—the myriad-pointed apex of this continental pyramid.

Below, the slopes are nobly clothed with vast and ancient forests. Theirs are the infinite fingers of root and branch which weave the rivulets from the icecaps above into a thousand nameless streams and loose again in a hundred twisting rivers. Wood and water are thus the salient features of the landscape. But they do not distort the ultimate truths of rock and aridity that hold here as above and below.

Farther down, the core obtrudes again in the lower terraced steps of bare, eroded rock. Here, hewn out like great buttresses and carven like bas-reliefs on the walls, mesa and plateau repeat in motif the pattern of the whole.

Thus it stands on the floor of the desert, deeply scored below sea level to show it anchored to the bedrock of a continent: the Colorado Pyramid whose lower section is composed of the terraced plateaus, and whose upper section is the forested mountains supporting in turn and as a multi-pointed apex, the bare lofty peaks above timberline.

It has always seemed strange to me that we associate the pyramidal pattern with the far wasteland of Egypt rather than with the New World where it most truly belongs. Here, as architecture, symbol and living form, the pyramid reaches its fullest expression.

In the Mayan, Toltec and Aztec civilizations it was a basic form. The Pyramids of the Sun and Moon at Teotihuacan, and the great Pyramid of Quetzalcoatl near Cholula with a base a thousand feet square, exceed in area the largest in Egypt. When the first Spaniards arrived in Mexico they found Aztec priests still conducting their great sacrificial rites on top of huge pyramids. Today, in remote areas of the Sierra Madres, one stumbles upon still more of them, like the small, ancient Pyramid of Tepozteco on whose top fifteen years ago I found bunches of fresh flowers.

In the Colorado River basin the prehistoric mesa-top pueblos were pyramidal in shape, as are the great pueblos today along the upper Rio Grande with their terraced walls reaching a height of seven stories. An Indian sitting wrapped in his blanket under his high-peaked Stetson—he too mirrors with his living form the greater shapes of land and man that have forever loomed upon his horizon. The Masonic emblem of the pyramid had a symbolic meaning to the white founders of this country as well as to the indigenous people of the New World. It is wholly appropriate, then, that the symbol on one side of the Great Seal of the United States is a pyramid. For all of these the Colorado Pyramid, carved out of the continent, might well be the mother shape, the ancient and eternal prototype.

In still another respect the Colorado basin offers a peculiar paradox. It contains one of the greatest river systems in America; and at the same time this whole vast drainage basin coincides exactly with the Great American Desert once officially defined as that "arid region requiring guides to watering places," or water holes.

That part of it now known as Wyoming Daniel Webster declared "not worth a cent," being "a region of savages, wild beasts, shifting sands, whirlwinds of dust, cactus and prairie dogs." Nevada today answers to the same description. Of Utah's 84,990 square miles, only 2,806 are water surface and the rest is a mile-high, desert upland. Arizona is a desert; less than 1 per cent of its 72,000,000 acres are cultivated and 90 per cent of this by artificial irrigation; and over fifteen years ago all the available water for this, except the Colorado River, had been put to use. New Mexico has the smallest water surface of any state in the Union.

THE COLORADO PYRAMID 7

Mountainous Colorado, the mother of rivers, twice the area of England and the seventh largest state, significantly ranks seventh from last in water area.

It is all an arid upland where water is scarce and precious—the lifeblood of its peoples. Rain here is not a casual atmospheric condition; it is a gift of the gods. The mean annual rainfall of the whole area is but a scant 10 inches. Even in the high snowy mountains of Colorado it does not exceed 16 inches, and in the Colorado Desert it averages less than 3. Little wonder, then, that upon rain has been built an ancient and elaborate religious ritual that was already formalized when Marco Polo went to China, and is still observed without change. Men here for centuries have danced for rain in prayers of moving rhythm.

When it does come it is a deluge: a cloudburst that floods a dry creek bed within an hour, claws and rends the land, and vanishes within the next. A mere average runoff of less than $1\frac{1}{2}$ inches, at that. Over such a maze of washes, gullies, barrancas, arroyos and creek beds, dry for most of the year and possibly for several years, bridges are impossible. One crosses these "dry washes" by what is known in Arizona as Wyoming Crossings and in Wyoming as Arizona Crossings—a slab of concrete laid across the bottom as a mere protection against quicksand. With water—more than with any other of the elements—man here is most concerned. Not because it is so abundant, but because it is so scarce and fickle. From prehistoric times it has been the one single fact that has influenced most the lives of its dependents. The belief that the country is steadily becoming drier persistently gains more adherents year by year. So that water—the conservation of it, the full utilization of it—is for the technological future, as for the prehistoric past, the greatest problem of man.

Such is the peculiar paradox that in this Great American Desert originate the greatest river systems in America.

From the glaciers, icecaps and snowdrifts on the peak of the Colorado Pyramid trickle the rivulets which form the headwaters of the Missouri-Mississippi, the longest river in the world, flowing 4,221 miles east and south into the Gulf of Mexico; the Columbia, which drains westward into the Pacific; and the Colorado, which cuts its way south to the Gulf of California.

Within a radius of fifty miles from the Grand Teton and the Wind River Mountains of Wyoming rises each of their main tributaries—the Snake, main tributary of the Columbia; the Yellowstone, of the Missouri; and the Green, of the Colorado.

Farther down in the snowy Sawatch Mountains of Colorado there is another great divide where the South Platte and the Arkansas curve north and south around Pikes Peak and flow eastward across the Great Plains.

Still farther down the continental watershed there is yet another where the Rio Grande rises to flow south into the Gulf of Mexico.

Thus the Colorado Pyramid: a vast desert upland and at the same time a mother of rivers—rivers that seep and trickle, run, rush, cut, slash and tumble down its slopes in all directions.

The river system that drains this tremendous pyramid itself is both immense and complex. It includes at least fifty rivers and as many more named creeks and washes. Any one of the Colorado's five major tributaries exceeds in size most of the other rivers in the United States. In such a vast network the majestic little Hudson would water scarcely a twenty-fifth of the arid basin; and all the storied little creeks like the Suwannee and the Wabash would be lost in the maze of cañons.

The Colorado river system is at once an international headache, a geographic skeleton, a hydrographic puzzle, a roll call of the most familiar names in the whole Southwest, and a symphony complete from the tiniest high pizzicato of snow-water strings to the tremendous bass of thunderous cataracts reverberating in deep cañons. At best we can here only distinguish its outline, call off its names . . .

The *Colorado River* rises in the Rocky Mountains of north-central Colorado. As it begins its journey down the western slope of the Continental Divide and westward into Utah, it gathers many more: the *Williams River* from the Park Range, and from the Sawatch the *Blue,* the *Eagle,* the *Roaring Fork* and *Frying Pan.* All are white-water streams frothing against the rocks and subsiding in deep cold pools in the shadows of snowy peaks, the greatest trout streams in America.

The *Gunnison River* is its main tributary in Colorado. It rises in the Sawatch Mountains near Monarch Pass, picks up the *Uncompahgre River* from the San Juans, and meets the Colorado near the Utah line at Grand Junction. The name of the junction—as that of the great cañon farther down—possibly derives from the fact that the Colorado was long known as the Grand.

Farther down, just above Moab, Utah, it is joined by two more rivers from the San Juans: the *Dolores* and *San Miguel*.

Meanwhile the Colorado's longest tributary, the *Green,* has been cutting a devious route through three states to meet it. The *Green* rises in the Wind River Mountains of Wyoming and is fed by a dozen creeks from the Grand Teton and the Gros Ventre Range to the west. From Wyoming it flows across the northeast corner of Utah, receiving still more melted snows from the Uintah Mountains, the only major east-west range in the Rocky Mountain system and the highest point in Utah. Looping through northwest Colorado, the *Green* is joined by the *Little Snake* from the Medicine Bow Range of Wyoming, and the *Yampa* and the *White* of Colorado. Curving back into Utah again, the *Green* gathers from the west the *Duchesne, Price, San Rafael* and *Dirty Devil* rivers of Utah.

Now, at the entrance of Cataract Cañon, just below Moab, Utah, meet the *Green* and the *Colorado*. At the end of the cañon, the swollen Colorado receives the *Fremont* and *Escalante* draining the great arid Kaiparowits Plateau.

Here in southern Utah just above the Arizona line, and not far below the Colorado's junction with its longest tributary, the *Green*, it meets its largest tributary, the *San Juan*, which also has traveled a devious route through three states.

The *San Juan* is not only the Colorado's largest tributary, but the largest river in New Mexico. Its annual discharge of 2,500,000 acre-feet [1] is over twice that of the noted Rio Grande. Yet it remains one of the least known rivers in America.

Named after St. John the Baptist, it too cries alone in the wilderness—one of the loneliest, wildest regions in America. The Navajo Indian reservation occupies a good half of the total area of its basin, and the Southern Ute reservation a portion of the remainder. Its headwaters are in the rugged San Juans of southwestern Colorado near Wolf Creek Pass. Flowing south and west it bends through New Mexico for a hundred miles, irrigating a mere 50,000 acres out of the 600,000 that could be used if the river could be utilized—little patches of corn and melons and small orchards, grown up in my own time beside the lonely trading posts that marked its route.

Its tributaries are many and wild, pouring from Mesa Verde, cutting deep gashes through the desert, stemming from a maze of impenetrable cañons.

The *Navajo, Piedra, La Plata, Los Piños* and *Mancos*

---

[1] One acre-foot of water is equivalent to the amount that will cover one acre with water one foot deep.

THE COLORADO PYRAMID

rivers, and the *Aztec, McElmo* and *Yellowjacket* creeks, all from Colorado.

The *Chaco, Cañon Blanco* and *Gallegos* rivers, with *Jaralosa, Carrizo* and *Largo* creeks, and *Deadman's Wash*, all from New Mexico.

And *Montezuma Creek* from Utah.

Swollen with all these, a savage red flood, the *San Juan* cuts its way through Utah unseen by man to join the *Colorado*.

In Arizona, just south of their junction, the *Colorado* receives the *Paria River* from Bryce Cañon, Utah. The mouth of the *Paria* is one of the most famous spots in the whole river basin, if not in the entire West. Here near Lee's Ferry is the only bridge across the Colorado south of Moab and north of the Grand Cañon, nearly a thousand miles. For almost four centuries it has been the rendezvous of Spanish, Mormon, outlaw, trapper and traveler alike, the 42nd and Broadway of this vast wilderness.

It also marks the division between the upper and lower sections of the Colorado Pyramid: the pointed peaks superimposed upon the forested mountains, above; and the plateaus falling away to the desert floor, below. It is the point of demarcation of the vast upper basin draining Colorado, Wyoming, Utah and New Mexico. Here, one mile below the mouth of the *Paria*, all the waters of the entire river system within its upper basin unite to form a single stream where the flow can be measured.

For now at the mouth of the *Little Colorado*, just below at the entrance of the Grand Cañon series, begins the lower basin including all the river drainage from Arizona, Nevada and California.

The *Little Colorado* rises in the Sierra Blanca Range of Arizona. Flowing north to the Grand Cañon it gathers the *Zuni River* from New Mexico, the *Rio Puerco* and *Lithodendron Creek* on whose banks is the Petrified Forest.

Shut in now for nearly three hundred miles by the Grand Cañon, the Colorado receives only two minor streams —*Kanab Creek* from the north and *Havasu Creek* from the south.

Below the Grand Cañon, at Grand Wash, the river turns a sharp right angle and flows south again, forming the boundary between Arizona and Nevada. Here it picks up the *Virgin River* from Zion Cañon, Utah, which has cut across the corner of Arizona to drain a total of 11,000 square miles, and *Meadow Valley Wash* from Nevada.

Farther south the Colorado forms the boundary between Arizona and California, receiving the *Bill Williams River*, which drains 5,400 square miles in Arizona. Red, gutted with sediment torn out of the Cañon, it crawls like a gorged snake under a brassy sky—a desert river. But with one more meal yet to swallow.

The *Gila River* joining the Colorado at Yuma is Arizona's biggest river. It drains an area of 56,000 square miles, part of which is in southwest New Mexico where the river first forms in the Mogollon Mountains. Then fed by a dozen tributaries it flows across all of Arizona to the California boundary.

The *Salt River* is its main tributary, a major river itself formed by the union of the *Black* and *White* rivers in the Mogollon and Black Mountains, and later fed by *Tonto Creek* from the high Tonto Basin.

Farther west, the *Gila* receives the *Verde River* and the *Hassayampa* from the north, and the *San Pedro* from the south. And still more—the *San Francisco, Santa Cruz* and *Agua Fria*.

One vast red flood pouring into another, the junction of the Gila and the Colorado completes on the desert floor the lower drainage basin of the Colorado Pyramid. Together they pour still farther south, like an immense turgid lava flow, through the delta region. For 17 miles the *Colorado*

THE COLORADO PYRAMID

forms the boundary between Arizona and Mexico; for another 80 miles it flows through Mexico, separating Sonora on the mainland from the upper district of the peninsula of Lower California. Winding through a maze of salty marshes, desert, mud geysers and dry lakes, split into the main channel and *Hardy's Colorado,* lately interchanged, the river at last moves slowly out into the Gulf of California and sinks its red waters into the blue.

Out of this vast network of rivers emerges the Colorado itself. Dominating land and man, it is the greatest single fact within an area of nearly a quarter million square miles. Bigger than its statistics, it is one of the great rivers of the world:

1,700 miles long, it is cliff-bound nine-tenths of the way and travels 1,000 miles through deep cañons. Its highest sheer walls are 4,500 feet—so high that falling snow never reaches the surface of the river. Its longest stretch of unbroken cañon walls is 283 miles, the Marble and Grand Cañon series. In its journey it drops a good two and a half miles from above 14,000 feet to 248 feet below sea level.

Aside from the diverse characteristics of its immense drainage basin and its own channel, the river itself is peculiar. Its volume is unpredictable. Averaging 20,500,000 acre-feet annually, in a single year it can crawl past Yuma with a mere 3,000 second-foot [1] discharge or roar by during the spring flood at the record rate of 380,000 second-feet.[1] Year by year its periods of high and low flood vary, always in mysterious cycles, but uncertain in length and magnitude.

It seems more solid than fluid. For one thing, its content is high in mineral salts gouged from the depths of the earth —carbonates, sulphates, and chlorides of calcium, magnesium, sodium and potassium. So high in fact that it was long

---

[1] One second-foot of water is equivalent to a flow of one cubic foot of water per second.

questioned whether it could be safely used for drinking water or even for irrigation, should the river be so utilized.

Few rivers, too, are so choked with silt. With 0.62 per cent average silt content by volume, it carries each year enough silt to cover 105,000 acres one foot deep. With this it has not only built a delta across the upper end of the peninsula, cutting off the gulf, but raising its own channel higher each year, the river now perversely travels—free of its confining cañons—on a bed elevated higher than the surrounding desert.

Yet there is plenty of silt left to deposit at its mouth —the equivalent of a cubic mile each century. Hurling this tremendous load into the Gulf of California, the river creates another phenomenon seldom seen anywhere else in the world—the bores, that conflict of the descending river with the force of the ascending tides of the gulf. And once its load is emptied, still another phenomenon takes place. Those tightly packed, cubic-mile blocks of dead weight accumulated for centuries suddenly give way. The whole skeletal framework of the earth is wrenched, distortions and displacements take place, and the reverberating shock ripples upward to shake the far mountains.

And this at last seems the ultimate task to which the Colorado is appointed by the Great Architect: to move bodily, sand by sand and peak by peak, through the measureless millenniums man calls eternity, the whole great Colorado Pyramid out into the sea. In this eternal palingenesis by which continents themselves are forever laid to rest and reborn anew out of their dead selves, nothing is static, nothing is still. Not even the great pyramid of the Colorado. Everything is alive, dynamic with constant change. Even the stones breathe; water is electric; the air is luminous with the memory of belching volcanoes.

We know so little of the earth we tread, on which we are such recent strangers! Only the river, the great red river, as the instrument in the hand of change, has written its ultimate meaning on the vast palimpsest of the living land. Here, if only for an instant, we can sometimes read it, still engraven on the pages of its uncovered terraces.

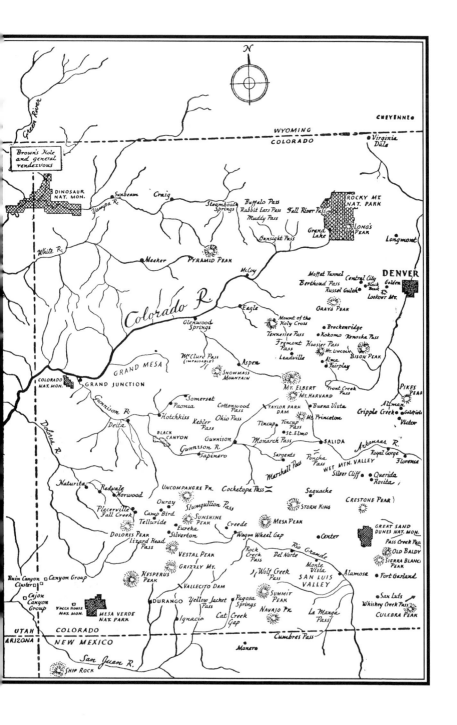

CHAPTER 2

High Country

NEXT TIME, by hook or crook, make sure you're born with a mountain in the front yard. It comes in mighty handy all the way around.

When you're no bigger than *that* you can hang on to the grimy window curtains and watch it hour after hour. Then you know it best with all its moods and mutations, its sternness, dignity and immeasurable depth. It is like the face of an old medicine man, which only a child can understand. Other times it's just a grand spectacle of a thing—a whole show in itself. In the evening when the sun snags on the rimrock and the hollows fill up with red and lilac, damson blue and purple; when the summer storms explode against its shoulder like soap bubbles filled with father's pipe smoke, and the deer-horn lightning sprouts from the crags; or even in winter when its slopes turn slick and green as Blue Ribbon bottle glass.

But from the day you start to school it begins to be useful. The mountain has drawn back out of your yard, receded across the railroad tracks, even a mite farther. But like you it's got bigger too; you can see it even from behind the schoolhouse privy, so there's no need to worry about losing your way home. Just keep it on the right, on the side where you carry your back-pocket handkerchief. That's the rule to follow when you start camping out in the lower cañons; climb a tree if you can't see it. But besides being a compass the peak is a timekeeper and weather prophet too. A peek out the kitchen window to see what kind of clouds hang over the summit is the way folks always start the day. Later on, when you fancy yourself getting along a bit better than your neighbors, it's something to measure your success by; the peak is pretty chesty itself. Toward the last, of course, it's the best of all. When everything else seems gone and you're just another old fellow sitting alone on the stoop with your pipe, its big cone helps to fill up the heart's emptiness and you know it's one thing that won't pass.

A mountain peak, all in all, is about the handiest thing to have around and strike up friends with. Our mountain was a whopper, a beaut of a peak. We got along fine.

It poured gold into our laps, tons of it. In fact it built the town, both Millionaires' Row and the shanties of Poverty Row across the railroad tracks. It lent a dignity to all our lives, brought into them an enduring sense of mystery, and whetted our appetites with the sharp winds blowing down the Pass—which was no blessing to be sneezed at either, considering the hundreds of poor, sick, half-starved rich people who came to it for just that purpose.

Another thing, you didn't have to travel with the mountain so close. It was the whole world heaped up in layers.

There was the sandy, flat mesa, the first stairstep from the plains, with its garden of queerly eroded red rocks and its dust storms. Halfway up the peak at almost eleven thousand feet was timberline, drawn straight across it as if by a ruler. It separated two more distinct worlds.

Below it to the mesa was a dark and blue world: the forest world of leaf and needle with its silent shadows and singing streams, its chattering squirrels and gossiping magpies, its timid deer and shy brown bear. But like obedient citizens of that still lower world below tide level, none of them ventured upward. No stout exploring fir or spruce, not a single wind-stunted pine, scarcely a blade of grass. Only a mountain sheep sometimes stood poised on a crag, but for no longer than a seal rests on a rock before plunging back down into the green.

Above was a barren world of pink-gray stone—and man, that unbelievable creature to whom no domain is forbidden, even that of the spirit. This was the upper half of the peak that gave life and color to our town.

To reach it one rode the little Short Line that crawled a mile up and eighteen miles around the south slope of the peak. You knew you were getting close when a tourist's nose started bleeding and the ladies drew back dizzy from the windows. Suddenly there it was, unbelievable world: a half dozen little sprawling camps, small clumps of shacks and cabins that seemed to have been dropped out of the sky into the gulches, and everywhere, crowding the streets and littering the hillsides between them, the vast gray dumps and stark gallows frames of the mines. All hanging to an immense slope of rock seamed with great gulches.

Nothing could have been livelier, more entrancing and more pregnant with anticipation. Hills and gulches swarmed with men, burros and machinery. The birth of a monstrous mystery seemed imminent at every moment. Yet one famous and hasty visitor reported it a scene of appalling desolation, adding casually that no cat could live so high.

It was a challenging indictment. Timberline became Catline overnight, and every indignant miner's wife made haste to disprove it. Henceforth our lives were overrun with cats. By dozens they were sent up on the daily train. They roamed the streets and dumps like burros, made the nights hideous with noise, and finally went wild in the lower gulches. Unfortunately the visitor who had instigated the deluge had tarried only long enough to interview a certain

Madam on the busiest street. To get even with him, the Town Council promptly renamed this, our local lane of prostitution, in his honor: "Julian Street."

It is a pity that Mr. Street did not remain longer. For the high country above timberline is like no other, and this was its boundary. In the absence of trees and foliage he would have seen here for the first time how everything stood out lucid and stark naked in new perspective. Like the weathered houses, the peeling log cabins, people's lives were stripped of decoration and verbiage by wind and weather. He would have noticed how everything tended to reduce down to stone. The faces of the men coming off shift were gray and colorless. The precious baubles in the parlor were heavy lead-gray samples, not needing roasting to bring out their hidden colors to the appreciative eye. Even the graves were blasted out of the hard granite with dynamite, that man could return not to dust but to stone.

A world, a life which even then seemed in the process of petrification. But still containing inside a hidden warmth. So watching on a winter's dawn the long queue of men plodding up the trail to the portal of a mine, a hasty stranger would be impressed only by the dominant drabness. He will not wait to see, a few minutes later, the window geraniums in their tin cans glowing like tiny fires lighted by the sun. He will not smell the yeasty buckwheat batter in the buckets set out after breakfast on every back stoop down the row of shanties. Nor see inside of an evening the walls livid pink in lamplight with the front sheets of the Denver Post papering the walls against cold. The tall peak itself is the core of an extinct volcano, and the running fire congealed in its veins still reflected the heat of desire which drew men here.

Everything, it seemed then, came from these high wastes of barren stone. I remember being taken by Jake to an impressive evening's entertainment. We sat next to the plank runway built down the center aisle from the stage, so that Jake in high humor and with boastful familiarity could snap the garters of the beautiful dancing ladies who tripped past. One garter I remember well—and also my unfulfilled ambition to grow big and bold and rich enough to snap it myself. It was a flower that came from no mere vegetable garden, a constellation that had never shone in any midnight sky. It was a beribboned rosette as big as Jake's fist and full of rhinestones and rubies and sapphires and diamonds. It was the most beautiful and vulgar, the most exquisite bad taste, the—"But, Jake, where did it come from? I never seen—"

The huge black-stockinged leg raised and pivoted over our heads, graceful as the steel arm of a crane. There was the close flash of light, color, ribbon and lace, the lusty wink of a dimpled knee, a raucous laugh. Jake made an ineffectual grab, and sank surlily back; it was a man behind him who was allowed to be successful. Girl and garter swept past.

"From the fourth level of the *Sylvanite!* Where the hell do you think it came from?" But Jake's habitual good temper was curiously out of whack; he might have spoken wildly. For certainly neither the fifth nor the sixth level of our mine ever produced anything to compare with it.

But all this, camp and people, still straddled timberline. It was still linked to the lower forested cañons, to the town below.

Our family folly lay even higher. Grandfather called it a mine though he had developed too many good ones to thus misname it. You could see it from the high saddle above the camp, a mere black speck far up the gulch. A bird could wing there in ten minutes. It took a boy on a burro an hour and a half. But if he waited at the station and rode up in the wagon with the supplies it was three hours up.

The speck had grown by the time he got there. There was the great gallows frame bestriding the shaft, a stout log cabin, and a small lean-to and corral of weathered aspen poles. And every great squared timber, every seasoned log and green pole, every stick of firewood, every nail—to say nothing of winch, steel cable and tools for the mine—had been snaked up here by mule and burro. One felt, upon looking downward upon camp and the dark slopes dropping from it, like a swallow resting on the side of a cliff.

Looking upward was worse. You were confronted with a steep terrain of bare frost-shattered granite that rose upward over two thousand feet. Nothing could convince the boy that this was the majestic, familiar peak he had seen from below since childhood. It lacked shape, outline and character. He simply felt imprisoned between sky and earth in a waste of stone. The feeling grew more oppressive daily.

There was no sound of rushing water, nor the sough of wind through pines. There was no smell of living green. No color relieved the surrounding monotone of rock. Only occasionally from deep underground rose a muffled roar, and at night there were new samples to assay. Here in this solitary little world of Grandfather, Abe and Jake, one lived a life of stone.

One blasted it, worked it, dug its dust out of ears and eyes. One cursed it, blessed it, prayed for it. It crawled into the blankets and into the dreams of night. It was at once the bane of your life and a resplendent future that at any crosscut might come to pass. For all that it was merely dead stone.

Then suddenly it happened, the boy did not know how. All this dead stone became intensely, vibrantly alive. Playing on the dump one morning after he had washed the breakfast dishes, he happened to pick up a pinch. In the bright sunlight he saw with microscopic clarity the infinitesimal shapes and colors, the monstrous and miraculous complexity of that single thimbleful of sand. In that instant the world about him took on a new, great and terrifying meaning. Every stone, every enormous boulder fitted into a close-knit unity similar to the one in his sweaty palm. For the first time he saw their own queer and individual shapes, their subtle colors, knew their textures, felt their weight, their strain and stress. It was as if in one instant the whole mountain had become alive and known.

No longer did the nightly assays seem meaningless chores carried out by a perpetually disappointed, white-haired old man and two gaunt, patient helpers. They took on the mystery of rituals that constantly attempted to evoke some deep hidden life within the stone. Each sample brought up from below, the boy eagerly rushed to examine. How did it differ from those on top—these cold, wet pieces wrenched from deep in the bowels of the peak?

"By cracky, the boy's got the makin' of a hard-rock miner!" grinned Jake.

"Dom fool!" snorted Grandfather. "What's he expect —nuggets as big as marbles?"

Late at night, tossing sleeplessly in his bunk, the boy kept wondering. The mountain was not one great big solid rock as it appeared from below. It was a million, trillion pieces all held together without cement: some hard, some soft, in all shapes and patterns, burned brown on the outside and gray inside, some with a purplish streak, but all with a preponderant delicate pink tinge against the snow. But it had lost its benign personality. It reflected a monstrous, impersonal force that pressed him from all sides. He was suddenly, mightily afraid.

"What keeps the Peak from fallin' down on us?" he blurted out in darkness. "I mean—"

From Abe's bunk came the usual silence. Jake let out another snore. But suddenly from across the room came two testy words in answer. "Isostatic equilibrium!" And then a moment later, "God Almighty, this time of night!"

Isostatic equilibrium: [1] it haunted him for years, both the phrase and its ultimate meaning. And not until long afterward did he realize that each of us has his own vocabulary for even Him who made the Word.

Thus he came to know that high realm of rock, the peak itself. Week after week the snowcap steadily receded. By day the drifts melted and trickled down into the cracks and crevices. By night the water froze and wedged the rocks apart. One heard, if only in his imagination, an eternity of sharp reports and booming explosions when the boulders finally split asunder. But to all this expansion and contraction, the rhythmic pulse of constant change, the peak remained immutable, bigger than the sum of its parts.

There came the day when the boy stood on its summit. For only a few moments the foothills, mesas, and the vast flat plains spread out 7,000 feet below to orient him to faraway Kansas and the Universe. Then mists swept up to cloak the forested cañons on all sides; it began to rain. He was standing—isostatic equilibrium to the contrary—on an island of rock floating in the sky. Far off toward Utah was another, and still another like a steppingstone into Wyoming. It was as it might have been at the Emergence. Or perhaps as it will be at the end of another paragraph in their geological biography when they will be the last to disappear. The feeling was indefinable. He did not know it, but it ex-

---

[1] *Isostatic equilibrium*

A high area, such as a mountain chain, is kept balanced on the earth's crust against a low area, such as a low plain or ocean basin, by "isostasy" (from the Greek meaning "equal standing").

According to the principles of isostasy, eroded surface material from the mountain is constantly being moved to the ocean basin. In time this loaded part of the crust sinks, forcing solid rock, which behaves as a plastic substance under great pressure, to flow horizontally back underneath the eroded mountain. Thus the two blocks, or the high area of erosion and the low area of sedimentation, are kept in proper balance or in isostatic equilibrium. The comparatively level plains area between them remains unaltered.

pressed at best the timelessness which is the core of that we know as time.

The Ghost Towns

Nowhere is expressed better this relationship of the perishable to the eternal, of man to mountain, than here on these high peaks. And nothing can be more weird than to ride suddenly down upon one of the empty hulls of our lusty youth. The stout chinked cabins strung out along the stream, the single street with its patches of wooden sidewalks, the false two-story "Leadville fronts" of hotel, store and saloon, even the dignity of a tiny church—how perfect and complete!

Then suddenly you notice all the chimneys are peculiarly empty of smoke. Not a lamp is lit against the lengthening shadow of the mountain wall. No men troop down from the workings above. No child hop-skips along the corduroy.

"Hul-lo-a!"

The rocky wall alone gives answer. And in eerie silence you ride through a town bereft of all life. It may well have been emptied just overnight. A saloon door bangs open, a bat flies out. Everything inside is in place, only the newspapers on the wall have yellowed; the girls' hats and tights seem somehow old-fashioned. An overturned whisky jigger holds a few rat droppings. A spider's web obscures your reflection in the mirror.

Outside the evening gathers. A coyote howls down the draw. A queer miasma hangs heavy in the air, compounded of silence and loneliness and defeat. You feel suddenly unclean as if in the presence of disease; frightened, as if by a mystery that will be forever unsolved. Clearly, it is time to get the hell out and make camp over the ridge!

No; nothing is quite so fabulous as these ghost towns of the Rockies. A man could spend a lifetime trailing down their sites, histories and legends. Everywhere they cling to these tall peaks: on the wind-swept summits, on the steep slopes, and in the lower gulches and cañons. Tin Cup, Tarryall, Hamilton and Querida are all deserted as Rosita, once a county seat. Nothing remains to mark the site of Buckskin Joe where a hunter's bullet, missing a deer, plowed open a discovery furrow in rich gold quartz. Nevadaville is deserted. Of Altman on the saddle above Bull Hill, named after my grandfather's friend Sam, and once the highest town in Colorado with a population of 2,000, there are left only a few stone foundations. Kokomo, eight feet higher and still the highest incorporated town, has a population of 44 old prospectors. Silver Cliff, once boasting two daily and three weekly newspapers, is gradually falling down on the heads of its remaining 200 residents. Gillette, where once bull fights à la Spain delighted hard-rock miners, has a population of ten. A dozen ramshackle buildings mark the site of Goldfield, a city of 3,000 when first I saw it. A few, like Alma below the mountain of Silverheels, named after a dancing girl with silver slippers, Granite and Sunshine, have been rebuilt—a board here, a shack there—without altering their appearance or character. Fairplay, center of one of the oldest and greatest mining districts, is now known only for its huge monument to the burro "Prunes" who worked so long and faithfully underground he went blind.

Huddled against Mount Princeton is a great hotel that evokes in every line the best of those lusty plug hat and gold nugget cuff-link days. Built over a natural hot spring, with a crystal chandelier hanging above a ballroom big enough to grace the Plaza in New York, it staggers woodenly aloft sprouting domes and cupolas, porches and balconies like a miner's drunken dream of the Riviera—a Monaco set down in a Colorado spruce clearing, as one of these camps was once described. Even a stagecoach petrifies outside, still waiting for the last guest to depart. Such are the remnants of a past unreal to us now and as completely vanished as that of a people whose remains an archaeologist might expose under the dust of a thousand years.

Up the cañon is the camp that gave it life: St. Elmo, named by a miner's wife after the only novel she had ever read. A train used to run up alongside the stream. Now the rails and ties have been removed, and if you're lucky you can crawl up there in a car. A magnificent setting, a town squeezed between two mountain walls one feels able to spit across. But of course unreal. It seems too much like a movie set, with a caretaker there to drag out an old ledger and show you the signature of Douglas Fairbanks, a foreign duke or two, and other famous visitors. Nothing can convince you he exists to sell stray fishermen a box of crackers and a can of Velvet. Year after year he waits for the company to return to location, for the reel to unwind again a life of extravagant dreams.

It is almost unbelievable how doggedly such old-timers hang on to this myth of a returning past. I can think of none more tragic than Mrs. Tabor. The last time I saw her, a lonely impoverished recluse, she was still living in the tool shanty of the Matchless Mine on Fryer Hill outside of Leadville. It was not yet spring up here in this once proud "City of Silver in a Sea of Silver" over two miles above tidewater, and the drifts stopped our car a half mile away.

Knee-deep in snow we broke trail up the slope. Fifty feet away we halted. The end of a gun had been stuck through a slit in the door. I continued on alone.

"Stop!" A thin sharp voice cut through the cold silence. "What do you want?"

"We want to see if Mrs. Tabor is still here. I think she'll remember us."

"Who's that woman?"

"That's my sister. She wanted me to leave Mrs. Tabor

a bag of oranges from California on our way down the pass."

Silence. The gun lowered. "She's kind of pretty. It goes with a good heart. . . . Are you sure she's your sister and not a newspaperwoman? That's all that come up here any more. There was that fellow who wrote that book about Mr. Tabor. When I wouldn't come out he stood up there on the ledge and threw big rocks down on the roof and said he'd bust it in. That's how he snapped that picture in the book when I come out!" The gun came up again. "Lyin', thievin' scoundrels, just like Bonfils and Tammen. None of them with the good simple heart of Mr. Tabor . . ." More crafty now: "No! Mrs. Tabor ain't here. She's to town. I'm just a friend of hers. What'd you say your names was?"

"It doesn't matter. You just tell Mrs. Tabor two friends came by, and I'll leave the oranges outside when I come back."

While Naomi, soaked to the knees, went back to the hotel to change shoes and stockings, I returned with a gunny sack half full of oranges to the bare snowy slope. To an abandoned shaft filled with water, at whose bottom Oscar Wilde had once popped champagne corks, and up which had come millions of dollars in cool shining silver—silver that had made this one of the most fabulous spots on the American continent, and this woman an international toast.

This time from inside the shanty came a low monotonous rumble—the sound of a recluse talking to herself, a sound more empty than silence. She must have heard me let down the sack on the stoop, and called out before I could get away.

"Wait!" The door creaked open a thin crack to let out the terrible musty cold of a fireless room, and with it the faint queer crackle of paper. She stood back of the door in boarded-up darkness, wrapped in rags and old newspapers, clutching the old rifle.

"Wait!" A sepulchral voice from a sepulchral past. "I can trust you, can't I? Is it—is things coming back, you reckon? Mr. Tabor told me, 'Hang on to the Matchless!' And I have. I am."

It was a question that contained its own answer. A perpetual question perpetually answered day upon night in the freezing silence, years on end. What man can subtract or add to such a faith?

A year later, to the month, she was found frozen to death inside.

But if the last old-timers are gradually vanishing, hundreds of burros, our "Pikes Peak Canaries" and "Colorado Nightingales," still remain.

So it was when I returned to our own home peak after a long absence. The Short Line had been replaced by an automobile road that few had reason to climb. The High Line and Low Line, electric tramways that whizzed miners around the dizzy curves between camps, had vanished without trace. Of the eleven camps themselves only two were left. Grandfather, Abe and Jake were gone and most of those who had known them. Our mine was an unmarked grave—one of a hundred dangerous holes in the hillsides. The gold was petering out at last—quite reasonably, considering that nearly $450,000,000 had been scooped out from an area of scarcely six square miles. Distinctly the cats were gone; Catline itself was a forgotten legend. But the freed, half-wild burros remained.

Long-eared mousy ghosts, they wandered the rocky hillsides scratching backs against the gallows frames of deserted mines. They crowded the waiting room of an abandoned railroad station to escape the stinging blizzards. Morning and evening they paraded up and down the street nosing like dogs at the back doors for food. And every day at four o'clock they were at the schoolhouse in a drove waiting to be ridden home. Outnumbering the few children three to one, they brayed and kicked and bit in rivalry for this remaining human contact. Those left, long ears flapping, dejectedly plodded homeward behind the rest. Mouse gray, slow and aimless, they too looked as if they were slowly turning to stone—the last to give in to these high peaks, more patient and enduring than even they.

There are fifty-one of these peaks over 14,000 feet high and fifteen hundred over 10,000 feet high in Colorado alone. With those in Wyoming, Utah, New Mexico and Arizona, they are the summits of the Rocky Mountains. They are the vertebrae of the backbone of a continent, the lofty ridgepole and the watershed of America, the Continental Divide.

Yet all this sprawling hinterland extending upward from timberline to the summits of these bare peaks is known simply as "High Country." No other name fits it so aptly. It is not a country that can be defined by the horizontal boundaries of state and man; its measurements are wholly vertical. It defies classification by name and range, and indeed most of its peak provinces are not yet named. It simply rears aloft, above plain, plateau and forest alike—a High Country of barren stone, the apex of the Colorado Pyramid.

CHAPTER 3

The Shining Mountains

THESE mountains have been many things to many men, but none have seen them whole. Like all great mysteries they are greater than the sum of their parts, and familiarity but defines better the terms of their enigma.

Young Zebulon Pike, impressed by their height, first glimpsed the peak that bears his name while far out on the Great Plains. Crawling slowly up the Arkansas he believed it a white cloud hovering on the horizon. Later he and his party made an attempt to climb it. It was still too high, he reported; only a bird could scale its lofty summit. The first Americans on the whole had an eye more for their material substance. They called these high mountains the "Stonies," and we still know them as the "Rockies."

Yet many other eyes have seen clearer their distinctive quality. From Cree Indians in northwestern Canada the French explorer, Pierre Gaultier de Varennes de la Vérendrye, wrote in his journal what he had heard of these mountains that seemed to glow by night: "dont la pierre luit jour et nuit"; and on his map he named the Rockies "Les montagnes de pierres brillantes." Meanwhile far to the south the earliest Spaniards similarly called them "The Shining Mountains."

The name goes back farther than the written word.

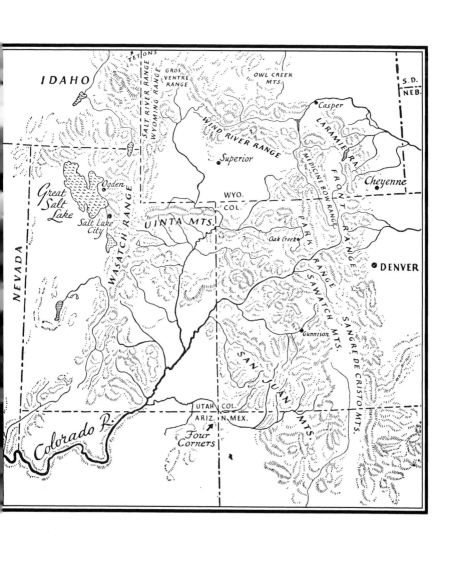

THE SHINING MOUNTAINS

Sacred Navajo chants of the Creation relate that at the time of man's Emergence, and long before the newly created world was lit by sun and moon and stars, these mountains shone with a strange luminous glow.

What it is no man can say. For here in the pitiless rarefaction no haze of dust discolors the perspective. There is no sunset. The sun drops white and sharp between the peaks, and the world turns abruptly dark save for the strange and shining effulgence of the mountains themselves. It is as if they are lit by a life within.

Indians believe in this subtle aliveness. Once, lying face down to recover breath after a steep climb, I felt the mountain stirring beneath me. Before I might have blurted out a foolish remark, one of the Indians beside me caught my eye. "She alive, all right, this mountain. She breathe too." Old Alfred Stieglitz would have understood what he meant, feeling the slow systolic and diastolic beat, the cosmic pulse within the stone. He once observed that the high shining towers of New York danced on moonlight nights. But the aliveness here is apperceived not only by an imagination sensitive and poetic as his. This sacred mountain, Indians say, is the center of our living world; and free from the vibrations of the jangling surface life below, with ear to ground, you can feel its deepest pulse.

We cannot dismiss these beliefs too readily. Medieval ecclesiastics preparing maps for the Crusades marked Jerusalem as the geometrical center, the navel of the world. According to modern geopolitical thinking, they were not far wrong. Balancing all the large land masses now known, in the concept of a world-island afloat in a sea which covers three-fourths of the area of the globe, we find that Jerusalem is indeed the center.

So too this mountain may indeed mark the center of gravity of the Colorado Pyramid. The Indians' tenure of it for sacred rites goes back to when it was the highest peak, though timberline has been modified hundreds of feet since and it is no longer that; certainly thousands of years.

How, then, describe these mountains? The Rockies, the America that we have always feared and always loved and have never known, and whose mystery abides in us forever. The great male Rockies.

One can as well attempt to make a blueprint of a cloud. Yet great in area as they are, a horseman can traverse all their zones of life in a single day simply by taking a trail upward. For here one mile in altitude is comparable to some eight hundred miles of latitude. Thus, not long ago, I rode with a visiting young professor and some Indian friends from the Sonoran Mesa of Mexico up through Canada to the rim of the Arctic Circle.

To the professional naturalist this was as lucrative a journey as it was to my hired friends and myself, mere lovers of nature. We started out one warm morning in early September. Burro and rabbit brush scraped our stirrups. Chokecherry bushes and hedges of wild plum were loaded with fruit. Soon the air became edged with the sharp clean smell of sage. There rose a low dark growth of piñon, juniper and Rocky Mountain red cedar. At the top of this rugged plateau we stopped before a typical New Mexican landscape.

A few massive, yellow-leaved cottonwoods marked the course of a deep arroyo. Along it stood a patch of corn—small stunted stalks of squaw corn with full and hardy ears. Behind it rose a wall of pines. Corn against pines! All bathed in bright yellow sunshine under a clear turquoise sky. There is no scene so common or beautiful. The distinctive coincidence and contrast of leaf and needle, of corn and pine, achieves here a perfect balance. It also marks another change of zones.

For now we left the warm, open world of yellow sunshine for the cool blue shadows of the forest. Squirrels chattered of our coming from the shaggy pines. A wild turkey blended silently into the bronze underbrush of scrub oak.

The trail narrowed, steepened. We wound upward single file, stopping frequently to rest the heaving horses. The yellow pines gave way to Douglas fir, lodgepole pine, Engelmann and blue spruce towering sixty, eighty feet high. A deep, ancient and mysterious forest, like a welling darkness through which only a trickle of light dropped from above.

At the bottom of the cañon far below, a stream poured whitely, feathering at numerous falls, then subsiding into still trout pools and beaver dams. Just past midday we stopped for lunch in an open glade. The sunlight seemed pale and weakened with alloy, and our fire felt comfortably warm.

Then, saddling again, we rode on. The great trees thinned out. And crossing a high divide we rode down upon a conflagration of color so astounding that many a stranger has mistaken it on first glimpse for the most terrifying sight in the mountains. But it was not a forest fire. It was a grove of aspens suddenly turned color by the first frost —a vast yellow glow with tongues of scarlet licking at the surrounding evergreens. Not until we rode closer could we distinguish the slender white trunks.

The quaking aspen is the ascetic of the forest. No other tree is so fragile in appearance, so delicate in coloring, and so susceptible to the slightest breeze. In spring the new small leaves are a bright electric green whose undersides soon turn a pale silver. Perpetually they shake in a nervous tremble. Yet like all ascetics the aspen has a remarkable endurance; not even a sturdy pine can stand its tremendous altitudes and forceful winds.

Appropriately enough, the Indians in one of their most subtle and delicate dances carry branches of the traditionally first-turned aspens brought down from these high mountain slopes. Scarcely a dozen old men filing out of the church into the setting sun of late September. Chanting softly, shaking gently the full-leaved branches in their gnarled dark hands. The Sunset Dance is really not a dance at all. It lasts scarcely ten minutes. But its truth is there

in the trembling leaves metamorphosed by frost to their gold butter yellow and deep orange as the setting sun, and in the bent bodies and trembling voices of their bearers facing also the sunset of their lives.

Of the hundreds of artists who have striven to catch in

paint every other feature of the mountain landscape, few have achieved the difficult feat of portraying aspens. It seems no coincidence that the most successful is he who also has best framed the Indian in his portraits. His illustrations here fulfill a long ambition of mine to have some bit of his work reproduced in one of my books. A native of the ancient Tatar capital of Kazan who had attained prominence in Imperial Russia long before he came here to live, Nicolai Fechin and his work need little commentary here.

The son of a local wood carver, gilder and maker of church images, he spent considerable time among the native Cheremis, Chiuvash and Mordva, still primitive and paganistic. During the Revolution he lived seven years on the edge of a great dark forest along the Volga, twenty versts from Kazan. Little wonder then that the Indians and the great wilderness of America have had for him a singular appeal. Here, more than in Russia and later in Bali and Java, he has done his best work.

A small, shy and unassuming man with high Mongolian cheekbones, Fechin is all of a piece. He has the simplicity of a peasant, an Oriental flair for color, a sympathetic insight, and an integrity as indomitable as it is rare. In his brilliant portraits all these qualities of a vigorous realistic tradition combine with a singularly direct and forceful style. Few men since Gaugin have had his courage of color, a feeling for the texture of sun-warmed skins. Under his broad touch the shadows darken among the pines, the waters sing. So that today his studies of the simple subjects found in this vast wilderness—an Indian ceremonial, an old prospector, a shaggy burro, a grove of aspens—are among the best that have been done. They are all portraits, with depth as well as surface. . . .

Swiftly now the timber shrank to a few grotesque dwarfs clinging with rheumatic arms to the crags. Mosses and lichens painted the somber rocks. Fresh snow lay in the shadows of frost-cracked boulders with patches of last year's ice. Climbing through the Upper Sonoran, Transition, Canadian and Hudsonian plant zones, we had reached timberline, the boundary of High Country, whose frozen tundra duplicates that of the arctic.

It was late afternoon and terribly cold. Clouds hung low over the bare peak ahead. We shrank back into the edge of timber to spend the night. By the time we had unsaddled, fed and watered the horses, and gathered wood, it was dark. Then after supper we climbed a ridge to see the moon rise upon a world whose silence has never been broken by a train whistle and seldom by an ax. An archaic world of mountains heaving, ridge on ridge, from the dark pelagic world below.

By foot or horseback is the best way to see the mountains, and it is the way I have always preferred to travel through them. When I was a child burros were plentiful. The last three blocks up Ruxton Avenue were lined solid with corrals, and all summer long the roads winding up the cañons were full of jogging parties. For the higher trails walking was no less a pleasure and less bothersome. Everyone wore hiking pants and boots, and spent Sunday in the mountains.

For the ailing and the idle there was the Wild Flower Special. The train consisted of about ten open coaches that left town every Sunday morning. The round-trip fare was a dollar and they were always packed. The stop was for two hours—enough time for a picnic lunch and to strip that particular meadow and its steep hillsides of wild flowers. The return trip was an unbelievable parade of color. Many is the time I watched from a high cliff that slow roofless train crawling through the gorge below like a long and solemn funeral cortege. All one could see was a bank of moving flowers: pale-blue columbines, yellow wild peas, mariposa and orange wood lilies, rose and cream meadow lotus, rare fringed gentians, purple violets, butterweed, harebells, forget-me-nots, asters, Indian paintbrush, anemones and roses. By the time they reached town almost all were wilted and promptly filled the ash cans at the station. No one minded. Next week there was a new stop, a different meadow just as lavish.

It was an era I remember with some nostalgia for its slow pulse and effortless charm. Free barbecues sponsored by Budweiser and Blue Ribbon. Wagon rides to gather wild raspberries and chokecherries for winter jam. Picnics in the mountains. Dances at Bruin's Inn. Catching trout at Buffalo Roost. Hunting arrowheads at Indian Rocks. Raymond and Whitcomb tours come to see the scenery. Cabins in the pines. The carriages, the hitching racks, the cast-iron watering troughs. Cutting the Christmas Tree. Steaks sizzling on the coals. Baskets of trout wrapped in green leaves. White water pouring over the falls . . .

Old Lady Gaines epitomized it best. A short dumpy figure with a face made swarthy by sixty years of assorted weather, she drove the oldest tourist hack in town. Every day she met the trains with raucous shout and waving whip. One noon just as a likely set of visitors were about to set foot in her hack, a brash newcomer whizzed up in his Ford.

"Step right in, folks! See the sights in an afternoon in solid comfort. Why eat dust behind a team of plugs?"

Old Lady Gaines wasted no words. She set upon him with her braided rawhide and lashed him within a scant inch of his life. Next day, to the amazement of the town, she was properly fined by the judge. At that moment the era ended. Trout retreated upstream. The streams dried up. The old bear who played havoc in the cemetery went over the hill. The elk climbed higher, the deer fled. New highways and Tin Lizzies had opened the mountains and a new era as well.

The Passes

With all their infinite variations the mountains comprise not only heaving waves of forests, but jutting cliffs, abysmal gorges and deep sunless cañons, vast open parks and tiny arctic meadows, small blue lakes, gushing warm geysers, mineral springs, cold trout pools, lacy falls, heavy cataracts and great soggy marshes, cones and craters of extinct volcanoes, bristling hogbacks, rolling hills of sage and cedar, high groves of aspen, immense flat-topped mesas, solitary bluffs and weirdly eroded buttes.

Yet of all these components perhaps the passes come most readily to mind. To all living creatures, even to birds susceptible to the air currents above them, they have been the immemorial gateways through the mountains.

Hannibal, marching elephants over the Swiss Alps (having only seventy peaks over 10,000 feet) into Italy, accomplished a major advertising stunt that built up a legend of spectacular, almost inaccessible European passes. A few later St. Bernard dogs built it still higher. Actually the famous passes through the Alps are only from four to eight thousand feet high.

Hannibal and the old Romans used Mont Genèvre, merely 6,102 feet high. The Mont Cenis, first mentioned in A.D. 756 and used in the time of Napoleon by the Russian army under Suvarov to cross Switzerland, is only 6,834 feet. Loible, Simplon and St. Gotthard are no higher. The big St. Bernard is 8,111 feet; and Col de la Seigne, one of the highest, is but 8,242 feet. Brenner Pass, late historic meeting place for two small men, is appropriately only 4,495 feet high . . . They are all scenic little arches compared to the high mountain passes over the Rockies.

A good run-of-the-mill pass here will only begin its climb from the summit of the Alps' highest, and their names are still wilder—Whiskey Creek, Wolf Creek, Rabbit Ears,

Slumgullion, Muddy, Tennessee, La Veta, Trout Creek, Kenosha, Mosca, Cumbres, Poncha and Soldiers Summit.

Marshall, the highest transcontinental railroad pass in North America—the route of the Denver and Rio Grande up from the Royal Gorge of the Arkansas, is 10,856 feet. Wolf Creek, six feet lower, is so steep and rugged that it defied highway engineers until 1916 and is still formidable.

Loveland, Berthoud and Monarch Pass are all over 11,000 feet. At the summit of Fremont, 11,320 feet, is the Climax mine producing 72 per cent of the world's molybdenum, and the highest post office in the United States. Yet nearby is Hoosier Pass, 11,542 feet, a mere rutted dirt road twisting between four peaks over 14,000 feet high and the old mining camps of Alma and Breckenridge.

Independence, the highest automobile pass in Colorado, is 12,095 feet. And Lake Creek Pass, still only a pack trail though once forced by stagecoaches and freight wagons, rises 12,226 feet high.

They are all incomparable and unpredictable; they span lofty horizons to link the old with the new; and they span as well some of the most spectacular space on the continent.

Through these defiles, first worn into ancient game trails by hoof and claw, have passed the fabulous parade of Indians, mountain men, twenty-mule team freighters, crack trains and boiling Fords that in one century has telescoped our entire history.

Ute Pass within my own time has encompassed it all. An immemorial Indian trail down to the medicinal springs at its foot, it was still used when I was a boy by the Utes who packed down it each summer. My mother's earliest tintypes showed it barely wide enough for the lumbering freight wagons which carried our men to seek fortune over the high horizon; and I still remember it as a wagon road to the new camp of Cripple Creek. Not until I was out of knee pants was it graded to permit passage in the first Fords. Still later the old Colorado Midland railroad tracks were removed, and just lately has it been made a new highway.

For nearly a hundred miles south the Front, or Rampart, Range of the Rockies runs like a sheer, insurmountable wall. Then there is another opening, La Veta Pass, on whose summit my grandfather erected the first building, a stone supply depot for teamsters freighting into the farther San Juans.

Ever westward, northward and southward, one wall after another heaves up on the horizon, their saw-tooth rims looming sharp and forbidding in the turquoise sky: the Grand Teton and the Wind River Range, Medicine Bow Range, the Park Range, Front, or Rampart, Range, the jagged Sawatch Mountains and beautiful Sangre de Cristo, the rugged San Juans, the Uintah Mountains—the only major range in the United States running east and west, the La Plata Mountains, the great Mogollons, the Black Range and the White . . . Seen anywhere all these mountains give one the single quick impression of being boxed in by lofty walls.

So it is that these high mountain passes take on a peculiar significance. Perhaps more than any other single physical aspect of the land, they are an index to a subtle aspect of our own character as well.

A stranger, helplessly dominated by space and mass or merely entranced by the view, will stare at the glistening, jagged peaks. But a native's glance forever ranges uneasily back and forth along the rim hunting for the notch that will let him through or over the wall. Gunsight Pass! It is a generic name for that feature of the landscape as ever-present in fact as in pulp-paper westerns, and there are few which have not been so known locally despite their other names.

Not always are they so marked. Some are barely distinguishable by a faint vertical line revealing where two ranges overlap like the backdrops on a stage. Others creep timidly into the dark lower cañons as if into a labyrinth. Still others strut boldly into the rising blue wave, and always it is a miracle like the Red Sea parting to feel the mountains draw back to let one pass.

But occasionally the gate swings shut. I remember being stopped one late afternoon in a little town at the foot of a pass into Colorado. A chain was stretched across the road and a long queue of cars was being turned back by the town constable. "Snow on the pass," he said. True, there had been a light spring flurry, the clouds still hung low over the ridges. But the lilacs were out, the wind was warm, and the pass was barely 9,000 feet high. Besides, this was May 7th. Imagine mere weather refuting the calendar in 1936! Good-naturedly we all turned back for an hour or two until the mistake should be rectified.

Instead, it got worse. The wind developed voice and temper. The clouds dropped like a blanket fallen off the closet shelf, and fresh snowflakes fluttered out like moths. Meanwhile more and more cars had driven in until the narrow street was jammed.

It was suddenly dark. Angry salesmen in light spring suits, stalled motorists, frightened tourists and perplexed campers made a dash to the hotel. Wives and husbands separated to crowd three and four to a room; cots were placed in the halls. When the hotel was full, citizens offered spare rooms and parlor sofas. By eight o'clock every café and lunch counter was out of fresh food. The proprietor simply waved cheerily at the cans on the shelf. "Point her out, mister. We'll open and warm her up." With astounding hospitality the doors of the Picture Palace were thrown open for free movies. As a last gesture a big fire was built in the lobby of the hotel, and the church organist was called to the piano to lead a community sing.

Outside, the wind had begun to howl.

By midnight, drawn together in the solidarity of those

who are called to share an unexpected and novel experience, we settled for the night. Fat businessmen wrapped themselves in overcoats on the lobby floor. A flour salesman passed round his bottle. Lights were snapped off. The fire died.

Yet no one slept. Upstairs a frightened child could not be hushed. Somewhere, in the church belfry perhaps, a loose bell was whipping back and forth, tolling faintly but wildly. Above all this rose the shrieks of the wind. Two state highway patrolmen came in, covered with snow, and tried to telephone. The line was dead. "What's up, officer? How long are we stuck?" asked a sleepy voice. There was no answer. As they went out, admitting a blast of cold wind, we caught in the glimmer of street lights a glimpse of icy pavement and whirling sleet.

Stark, unreasonable fear rose. What had happened? A snowstorm had blocked the road over the hill for a single night. It was more than that. It was as if the hairline balance of man's dominance over nature, achieved after centuries of struggle, had suddenly been lost and we were again at the mercy of those elements we had ignored so long and comfortably. Our thoughtless assumption of security had been destroyed. The smooth course of our lives had been rudely checked. In that small mountain town, high on the breast of a wild and lawless continent, we were as helpless, lost and frightened as passengers on a stricken ocean liner—and that loose bell, pealing wildly through the dark cañons and overhanging cliffs, carried the same tone of frantic futility. With its endless, unanswered ring in our ears we awaited dawn and then rushed to doors and windows.

Before us at the end of the block rose the great wall of Colorado mountains, two thousand feet high. It was white with snow. What we could see of the entrance to the pass was an unbroken drift. Branches of trees littered the icy street. Cloth tops of cars were shredded canvas.

The silence was suddenly broken. Three women rushing for the hall bathroom upstairs had got into a wrangle.

All morning a crowd of people thrown out of joint milled in the street. Townspeople, so hospitable the night before, were curt at answering questions. The price of tire chains doubled. The stark, unreasonable fear of night was replaced by a surly, unreasonable anger—at what, no one knew. They only felt that immense snowy wall of indomitable rock press in upon flimsy brick and shingle, upon minds and nerves.

At noon escape was offered. The road back across the plain was cleared; people were advised to take the long detour. But no one budged. American stubbornness set in. Humanity in its ceaseless struggle with nature cannot give in anywhere.

At last, near four o'clock, a single funereal line of cars struck up over the hill. The first two promptly slid off into a ditch and heeled over with broken axles. Damned by road crews, blocked by the snowplow, the rest of us reached the summit in low gear after five hours. Here we were forced to wait two more while another file followed up a second plow from the other side. There was no room to pass; the drifts on each side were higher than the tops of the stalled cars. By the time state highway patrolmen finally took hold, it was after midnight and bitterly cold. Fires were lit, and between these we snaked downward yards at a time with sleepless eyes and numbed hands and feet. And when at last we finally got down the pass in wan daylight, it was only to find the level highway still partially obstructed by fallen telephone poles and tangled wires.

Such may be the annoying hazards of a modern pass for travelers. But without these open passes the mountain walls close like the jaws of a trap. A whole people may be caught and held in helpless imprisonment.

The Spanish-Colonials

The remote little villages high in the Sangre de Cristo Mountains of New Mexico hold such a people. The same trap that a century or more later was to close upon the Anglo-

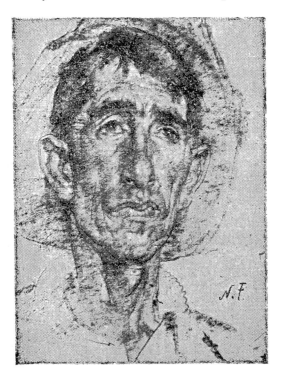

Saxon settlers in the mountains of Tennessee and Kentucky caught here the Spanish-Colonials, the first settlers in our America.

The violent rip tide of conquest receded. And stranded here for over three hundred years, like stagnant pools of culture, the little adobe villages lived on, isolated in remote valleys. They are the most fabulous villages in America. In

THE SHINING MOUNTAINS

them medieval Spain lives on, and the archaic syllables of Cervantes sound in their tiny plazas.

Still today they fringe one of the few, almost inaccessible wilderness areas remaining. A tortuous dirt road is their only link. Creeping through a dark virgin forest, it crosses narrow meadows between the lofty ridges, skirts tiny box cañons, lurches down into small valleys. In each of these clearings huddle the thick-walled adobes with their stout, high-pitched log roofs. There is a general store or two, several cantinas, a church, and a sprinkle of houses along road and stream. Perhaps only a dozen buildings. But a village as villages are in this remote highland.

Between them in summer travel old creaking wagons, jogging horsemen, groups of barefoot women trudging in the dust. It is the route too of cattle, sheep and goats led up to fatten on the tender, lush mountain grass. Jolted to pieces by protruding rocks, stalled by washouts and avalanches, a car has little business on this road. Deer bound across it, freezing in the glare of the headlights long enough to crumple a fender. A snowy arctic owl crashes through the windshield. Porcupines are forever getting under the wheels, piercing tread and inner tubes alike with their quills.

But in winter these tiny settlements are inaccessible by hoof and wheel, and to outsiders practically nonexistent. Each is enclosed by an impassable wall of white. Only then does one experience to the full that impact of the mountains unknown to prairie folk—their unconquerable resistance, their terrific pressure, and a sense of utter abandon.

One sees it reflected in the people. A strange ingrown remoteness in the depths of their unfathomable black eyes. The look of a people lost, forgotten and abandoned for over three centuries, imprisoned by cañon walls. For two years I lived in one of these villages. Thus does one learn why mountains forever remain the last areas to be permanently settled; why mountain folk are always the most sectional, aloof and obdurate to change. In the heart of both land and people there still remains a last wilderness which time and change have never opened.

Life is reduced to bare existence. There is no doctor, no priest. A native midwife or herb doctor breaks trail through the snow. Fresh food gives out quickly. Faces pinch and pale. Gradually the tension mounts. The whole village is epileptoidal. And the symptomatical spasms begin.

The name of the one of these villages means "rock," and it is the folk belief that every spasm, every quarrel, begins with a thrown rock. So now when a woman is accused of witchcraft neighbors stone her hut. An old feud breaks out again. A man's ear is clipped by a bullet one dawn as he breaks ice in the stream. There is a sudden stabbing at night in the plaza.

Over the tiny imprisoned pueblo the mountains keep rearing higher, whiter, bluer. And down upon the people bears heavier the latent violence, the insupportable silence. The epileptic tension grows. One knows that the crisis, the catastrophe will come.

Then suddenly toward spring it happens. One hears across the vegas the thin sharp whistle of a reed flute, the most eerie, fearful sound in the mountains. It is the screech of a Penitente pito, and it snaps instantly the long, mounting tension of winter. There begins a grim and fantastic Passion Play that has no equal on earth.

Men vanish from the plaza. The women, grim-lipped, sit quietly at home making panocha. A stranger asks no questions and wisely keeps safe within the village. These are secret rites that take place in the lonely dark cañons outside, and many profane visitors have been beaten and shot for their inquisitiveness. But after two years' waiting, word came for Fran Tinker and me to stand next afternoon under a certain great cottonwood at the mouth of a near-by cañon. We waited an hour more. At last, far off, sounded the pito and an ancient, almost Gregorian chant. Down the hill came the Hermanos Penitentes bearing a cross to lead us back to the massive, thick-walled, windowless morada. For us this tremendous Passion now became a naked and living reality.

In a hundred cañons it was taking place. Men fasting on panocha brought to the door of the morada secretly at night by the waiting women. Praying in the cold, candlelit darkness. Then marching out to the Calvario. The rezador leading the ancient chant. The pitero with his flute. And behind the Hermano Mayor, the Penitent Brothers stripped to the waist, chanting rythmically and whipping themselves with disciplinas of cactus.

On Good Friday, the hour that Christ died on the cross, it reaches the same bloody climax. For here too there is a Cristo staggering under the beams of a heavy cross. Behind him the whippers are more violent. In the morada the sangrador has gashed their backs. Four cuts, down and across in the shape of a cross, to allow the blood to flow freely without raising welts. So now at each sixth step, they bring up the braided cactus scourges over their shoulders and swing down upon their cut backs. The air resounds with soggy blows. The snowbanks crimson with spattered blood. Many of the penitents' heads are covered with black cotton hoods so that with due humility they may suffer anonymously their common sins. Sometimes the procession includes a "cart of death" with fixed wooden wheels which must be dragged by the Brothers. In it rides a skeleton draped in black, carrying a drawn bow and arrow. Here the trail is too narrow, the climb too steep. But the presence of death remains.

It is certain that in former times the Cristo was nailed to his cross. Today he is merely tied on with ropes drawn tight to stop the circulation of the blood, and left until he faints. Even so men have died on the cross, and their bloody shoes sent home as a sign of their complete expiation.

Then, haggard and tortured, the penitents return to the morada.

Here women and friends are waiting for the last and shattering act of the Passion—the tinieblas. How cold and dark and mysterious it is as one crouches on the bare hard floor. The twelve candles on the rude altar have been extinguished. The psalms are finished. Now as two thousand years ago there is darkness over all the land from the sixth hour to the ninth. And the multitude waits, one neighbor indistinguishable from another, breathing hoarsely with fear and cold.

Suddenly you hear the dragging steps, the tortured sobs outside. The great wooden doors creak open. Then, indeed, the veil of the temple is rent in twain, and the earth does quake. Here too Christ dies amid a wild pandemonium. Women break out in sobs and shrieks, rattling tin cans filled with pebbles, shaking rattles, beating on pots with pokers, on dishpans with old iron spoons. The amatrada—a noise-making machine of toothed wood—is set in motion. Somewhere a man drags chains across the floor. Another beats on a shovel. All in heavy darkness, amid shouts and shrieks and sobs.

Most observers, taken by the dramatic spectacle of the procession to Calvary, have ignored the tinieblas. Yet their effect is more shattering. They not only crack the eardrums but shake up into a new pattern the very corpuscles of the blood. All sense of the personal is obliterated. This is the resurrection in its most ancient meaning. Thus winter gives way to spring, with roaring avalanches of snow and rock. The passes are opened, and with them all men's hearts. The terrible tension is broken, its crimes and passions expiated. And with the flood of grief and pity and compassion, with tears, come the new spring rains. . . .

The Hermanos Penitentes—Penitent Brothers, or Los Hermanos de la Luz—Brothers of Light, derive from the

Third Order of St. Francis. Their rites were brought to New Mexico in 1598 by the six hundred colonists under Don Juan de Onate. Outlawed as a sect by the Catholic Church, the order perished elsewhere, even in Old Mexico. Only here it persisted. With its rites the Spanish-Colonial settlers remained devout when abandoned by the padres. Still later when new and greedy circuit-riding and parish priests robbed the people of the rites of baptism, marriage and death by exorbitant fees, they found it still deeper rooted. Fifty years ago the Church formally attempted to abolish it but with no success. The Penitentes number thousands, and extend throughout all northern New Mexico and into southern Colorado. They are a subtle force in politics and control village affairs. Almost every influential man helps to support a morada. Yet separated from the Catholic Church in spirit, the Penitentes are members in fact. They practice their own rites secretly through fear of excommunication; and the priests, knowing themselves powerless, remain conveniently blind to their activities. The village where I lived is their headquarters. Within a day's walk I could point out at least a dozen moradas. Yet all the time I was there I could not tell which of all my friends were Penitentes—or were not. I never asked. It was enough to go swimming with them in the presence of a stranger. Then they never took off their shirts, which would have revealed the telltale scars of the whips.

The roots of the Penitentes go deeper than any church, deeper than even their Christian origin. Part pagan, just as Indian ceremonialism is but a primitive faith covered now with a patina of Christianity, they reach down to the ancient sacrifice of blood. To the primitive acceptance of death as a part of life. Thus did the ancient Aztecs tear out human hearts on top of their lofty pyramids, and do the modern Mexicans celebrate the Day of the Dead as a national holiday.

But beyond all this the Penitentes are rooted in these mountains which alone have nourished and kept them alive. In the great fear and stark reality by which man confidently at last waters the earth with his blood and returns his flesh to dust that it may bear again the fruit of his likeness.

CHAPTER 4

Mesa and Plateau

FROM the high white peaks above timberline, through the forested blue mountains, the river cuts its way down through a great red plateau before reaching the tawny deserts below.

This rugged upland comprises the lower terraces of the Colorado Pyramid. It is a vast wind-swept expanse of sage generally bare of field and forest. A fantastic realm of rock slashed by immense cañons and gashed by washes, gulches and arroyos. Here wind and water, the immemorial forces of erosion, have carven their weirdest shapes.

Broad, flat-topped mesas float like islands in immeasurable space. Jagged red cliffs, tall graceful spires, ungainly mushrooms of sandstone, and squat ugly buttes stand like steles of a world ancient and forgotten. Whole forests lie level, petrified into stone. Upended walls of white limestone rise salmon pink out of the dawn and dissolve in the lilac dusk. Ridges of black volcanic tufa bare teeth to the cutting wind. All these shapes—of towers, embattlements, bridges, castles, ships, beasts, birds and seas and cities—the tumultuous white clouds forever strive to mirror in the turquoise sky. They will not be duplicated. Seemingly tenuous as the dust devils that glide spectrally across their far hori-

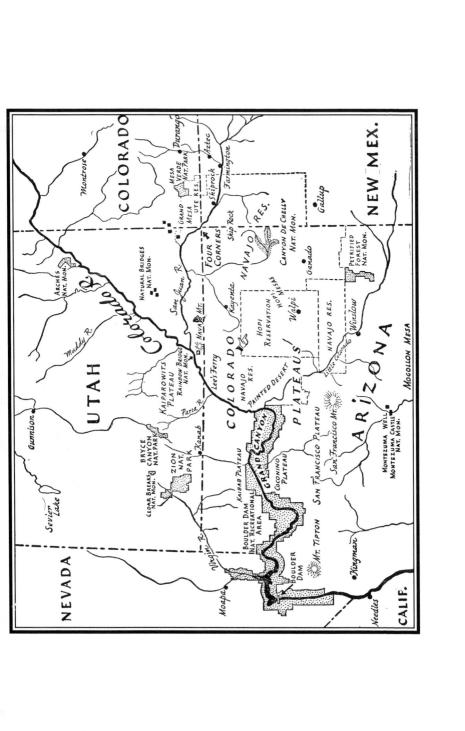

zons, they stand rooted and indestructible in sunlight and silence.

Unlike the peaks and the mountains, the plateaus have little meaning in height. They belong to the horizontal, and are measured only in time and breadth. This is a wide country. The clear, thin air magnifies and distorts still larger its dimensions, and only imagination can bound it.

As the Land of Standing Rocks is the name given the region for its weird shapes, so the Painted Desert identifies it by color. Its predominant note is red. The color for which the earliest Spaniards named the plateaus themselves, the river which drains them, and a portion of the whole region itself: Colorado, or red.

It is not altogether strange that one who knows this country well thinks of its four areas in terms of color: the white peaks, the blue mountains, the red plateaus, and the yellow deserts. To the Indians color always has been a geographic designation. Deriving perhaps from the color of each of their successive, mythical four worlds—the Dark World, the Blue World, the Yellow World, and the present White World—it carries a chronological meaning as well.

The Navajos still use colors to designate the four principal directions. The four sacred mountains marking the boundaries of their plateau homeland are similarly known by the colors out of which they were created: Blanca Peak to the east, made of white shell; Mount Taylor to the south, of turquoise; San Francisco Peak to the west, of yellow sand and abalone; and La Plata Range to the north, of jet and sand. The Pueblos, adding zenith and nadir to complete the ceremonial six directions, retain the same principal colors: white, blue, yellow, black, red and varicolored.

All this may not be so fanciful as it seems. Modern science, recognizing colors as merely certain wave lengths of light, takes cognizance of the substance that reflects them. So that this vast plateau of eroded rock, vibrating like the snowy peaks and forested mountains to their own keys, reflects that wave length of the spectrum we designate as red.

Not only in color does its distinctive vibratory key express itself, but in design and sound. Every schoolboy recalls the experiment in physics wherein the jangling of a key ring is translated into an electric current whose graphic sine curves are projected visibly on a screen. Walt Disney in a prologue to the movie *Fantasia* repeated the performance with various members of Leopold Stokowski's orchestra. You heard a few bars from the player, then saw the sound transcribed on the screen in its curious, inherent wave form. A simple and amusing display, it was amazing as well. For, while the notes of the flute appeared in forms rounded and Oriental in character, the beat of the drum built the same angular structural designs inherent to our Southwest plateaus. It was as if out of the drum you saw emerging not only the fantastic profiles of mesa and plateau, but the like patterns of Pueblo ceremonialism, Hopi pottery friezes, and Navajo blankets and sand paintings.

Not casually is a certain Navajo blanket designated as a "Two Gray Hills." Both patterns stem from the same source. From the vibratory wave length of the land derive ultimately the designs of its primitive people's art and architecture; and their music is pitched to the same frequency. Nowhere is this more true than here. No land so dominates its people, and no people are so sensitive to its basic vibration. For centuries they have been so attuned to its rhythm that their harmonic relationship with it constitutes, in the deepest sense, their religion. Conversely, they have given back life to the land. No other area in America is at once so barren and yet so vibrantly alive, attesting thus the reciprocity between land and man.

This, then, is also Indian Country—the traditional

homeland of America's only indigenous peoples, and the archaeological background of their prehistoric ancestors.

A passing stranger will not need to identify it on a map. He will recognize it at the first glance. Out the car window he will glimpse a lone and ragged Navajo horseman emerging from a sandy wash; and sixty, seventy miles away a solitary mesa gleaming red in the desolate plain. No more than this. For here as everywhere are integrated all the subtle and powerful qualities of this arid upland—its vast dimensions of space and time, its fantastic shapes, its color, and the lingering presence of its aboriginal inhabitants who still wander like living ghosts their native range.

The Woman Who Rode Away

Space, interminable, haunting, empty space forever and everywhere stretching toward infinity in the stark brutal clarity of the thin dry air.

That was all he saw between the ears of the plodding team as worn out by anticipation he slumped down in sleep. All that he saw next morning when he awakened in the trading post of Shallow Water. And all that he was to see for months thereafter.

For a boy always boxed in by lofty mountain walls it was at first a disturbing experience. He felt as naked and exposed as the land. But things straightened out like they always do. Instead of a stout log cabin there was the massive adobe trading post. Replacing Grandfather, Abe and Jake were Bruce and the Vrain Girls. It was as simple as that. And in this new little world island he soon became wholly engrossed.

Shallow Water was an L-shaped adobe fortress with walls three feet thick and iron bars over its small windows. One wing held our private living quarters. The other consisted mainly of an enormous public room flanked by a long counter, walled with shelves of groceries, hung with pelts and blankets, and cluttered by glass cases, an old potbellied stove and bursting gunny sacks of onions, pinto beans and corn meal. Here Bruce did his trading with the Indians who came plodding across the pelagic plain on wiry, shaggy-haired little ponies or in old creaking wagons. Shapeless moon-faced squaws in brilliant velveteen blouses and voluminous gingham petticoats. Slim-hipped, arrogant Navajos with bright headbands holding back their uncut hair. Somber children who squatted silently, hour after hour, in front of the candy case. Sometimes a party of Utes down from the mountains. Often a bunch of Apaches with long hair braids dangling from under their high-peaked Stetsons.

What did they all want? To buy sugar, salt, flour, a piece of hard candy, a bolt of gingham, a slab of the "white man's thin meat," a can of cold tomatoes. To sell a skin, pelt or blanket, wicker baskets, earthen pots, a piece of silver. To stand or squat hours on end rolling cigarettes or picking lice out of each other's hair. And to suggest delicately, if Bruce seemed in a humor, a drink of his free and foul Arbuckle coffee.

But mostly just to crowd there in the great room and mill around in the courtyard. For this was the focal point of nearly a thousand square miles and a thousand lives. Bruce knew it well. Game in one leg—soon to be paralyzed, he lounged quietly at ease behind the counter with eyes sharp as theirs and equal patience. A man feared and respected, he was master of a wilderness domain.

At last toward sunset the post cleared. The horsemen and the wagons drew away. A people like sheep threading the empty plain; vagrant as the stars threading the dark plain above. Bruce began to lock up his blankets and pawn silver in the rug room. For the first time he spoke to the boy and in English. "Suppose we better light the lamps and stir up a fire for the Girls, eh? Any water in the pot?"

Such was the boy's daily life, a monotone that was

never monotonous. His two old-maid aunts, the Vrain Girls, gave it added color. Only America could have produced their like. And in those days when convention bound women tighter than their whalebone corsets, they were no less than fabulous. Certainly they had seemed incongruous at home.

For there on rare occasions they returned to a ramshackle frame house beside the railroad tracks. Carrying the indefinable stamp of wind-swept space, they gave a singular impression. As if they dared not settle farther away from this ready escape back to their sandy wilderness. Even the old house seemed out of place. It was completely littered with Navajo blankets, Apache and Ute baskets, Hopi pottery, silver bridles, chunk turquoise and silver jewelry, all gifts given them on their successive departures. A trading post itself.

The Vrain Girls themselves were less impressive. They were sharp flint knives sheathed in a dark wrinkled leather. They had forgotten how to talk. The mothy, old-fashioned dresses taken out of the closets hung on their small wiry bodies with careless indifference. The hands that held their teacups were brown and calloused. The Girls, as the neighbors said, were just plain queer. No wonder! Imagine two respectable ladies goin' off to live, the Lord knows where, among a bunch of greasy heathens! And soon, without saying good-bye, they left again for some indeterminate place twenty miles by wagon from the end of the railroad.

Now here they were and the boy with them. Just why they were here he did not know. Some sort of Indian agents. All day or for days they would be gone in their wagon. Then, as tonight, they would return bustin' with news, gabby as magpies around the supper table.

"Mrs. Black Kettle had a baby. Boy. A big fellow. Strung her up with a rope. Four of 'em pulling down on her middle. She wouldn't lay down . . ."

"That smart girl of Hosteen Bega's. Right smart. I think

he'll let her be taken into the school. The feedin' up won't hurt her none either. . . . Did he come in with more pawn?"

Really it was Matie, the younger sister, who did most of the talking. Lew and Bruce just sat there listening. When they talked it was in a sort of code.

"There's some talk of a yegua parda around," said Lew at last. "A white stocking on the off foreleg."

Bruce shrugged. "I never pay any attention to strange horses," he said easily. "Never get outside."

The woman slid a heavy raindrop bracelet across the table to him. "Yellow Wolf's wife. I happened to stop at the hogan. He's visiting up in the Four Corners. Maybe the Monument."

Silently the boy deciphered. Yellow Wolf had been at the post all afternoon. Lounging in the doorway with usual nonchalance but with wary eyes. The buckskin was tethered down at the creek. He had been ridden hard; his coat where he had sweated under the saddle blanket was ridged and stiff. This in itself told half the story—that Yellow Wolf had stolen him and stopped by at the post for supplies. Three sacks of tobacco. That foretold a longer ride yet into wilder country, as far as the Four Corners or Monument Valley. His wife had sent back a piece of silver in payment, knowing he would stop and trusting both the trader and the white woman.

That was what their few words said. When the sheriff rode in a day or two later they wouldn't say that much.

Lew sat with her hands wrapped around the coffee cup. It was like a little pot of black earth. From it she watched grow a long gray stem with spreading branches.

"Our old friend finally went to Klah," she said after a time. "She had forty head of sheep."

"Plenty," answered Bruce tersely. "I have promised coffee."

The boy's eyes shone. This was momentous news! The

old woman had been ailing for years and saving for a healing ceremony. Now Klah, the medicine man, would call a great sing. A thousand Navajos camping on the plain. Dancing and singing all night for nine nights. Beautiful sand paintings in the new ceremonial hogan. Horse racing in the daytime. Much trading. And Bruce was giving coffee. That meant it would be close. Maybe it would be soon! Maybe . . .

The man observed his quick breathing and rapt attention without turning around. "Well, I reckon we'd better be washin' the dishes before the water gets cold, eh?"

Obediently the boy got up and started clearing the table. Suddenly he felt the trader strike him a quick sharp blow on the wrist. "Not till the thunder sleeps. But just a piece down the draw. You better keep this to wear to it."

The boy looked down. On his wrist was clamped the big turquoise-studded raindrop bracelet of Yellow Wolf's wife.

With singing heart he began scrubbing dishes. Matie and Bruce had begun their evening game of checkers. Matie giggled; she had got into his king row. Bruce puffed contentedly at his pipe. He had suddenly become white again, like Matie. All white.

But Lew still crouched as over her little medicine pot that had grown a flowered plant. Her eyes in their dark wrinkled sockets had a queer faraway look. Like the People she always seemed to be seeing things behind and beyond the shape of things, sort of. Suddenly the boy realized it. She was as queer here as the neighbors had said she was at home. She wasn't white. Not deep down inside she wasn't. And in a flashing, wordless revelation of feeling he knew why.

A year or two later, after he had returned home, it happened. Matie arrived alone and took up permanent residence in the ramshackle old house beside the railroad tracks. She seemed suddenly changed. Perhaps after twenty years of work and privation she had just naturally worn out. Eventually she succumbed to a late marriage and moved away again for good.

What had happened to her sister Lew she could not say. For months she resisted the probing of the family, and they the neighbors'. The boy could hear the whispered echo of their gossip, the sudden "Sshh!" as he entered the parlor upon their ceaseless conjectures. Gradually it began to grow into a strange family mystery, a tragic shame like that of a seduction or an illegitimate birth, which festered in the dark silence behind closed shutters and disgraced them all.

But finally, before Matie left, the boy learned what had happened to his other old-maid aunt. She, indeed, had committed the one crime society will not forgive. It seemed that Lew had grown steadily queerer. She resented the increasing influx of homesteaders, began to avoid all whites, and stayed longer and longer with remote groups of Navajos. Abruptly one morning she announced to Matie that she was leaving. The country was getting too crowded, she said; she was going downriver. Alone on horseback and carrying a single saddlebag, she rode away and had never been heard of since.

Again, as once before, an instant revelation of the meaning of her queerness comforted the boy. He remembered the vast empty plateau stretching away on all sides, the look in her eyes. Never in the years that followed did he assume for her the existence of a Navajo lover, considering her age and the taciturnity of the people to whom she fled. Wherever or however she died, the thought was not worrisome. No woman ever lived a fuller life or one within such great dimensions. Any old prospector or Navajo would understand the trail she followed. She was a woman on whom space had set its seal.

And gradually it swallowed Shallow Water itself. The boy never remembered exactly where it was. There remained only the vision of the post, and far away the muddy curve of river, a tiny town and the blue mountains of home rising beyond.

Vast and integrated as it is, the region has many names and variations of topography.

The Colorado Plateaus spread for some 45,000 square miles over all of northern Arizona. Between the Kaibab and Coconino plateaus the river cuts its deepest gash and the world's most stupendous chasm, the Grand Cañon. To the west, near the Nevada line, is the Shivwits Plateau. To the east is the wild, fantastically colored plateau known as the Painted Desert which often lends its name to the whole. Here lie the Petrified Forest and Dinosaur Cañon, with its footprints of the prehistoric monsters who stepped also into Utah, Wyoming and Colorado. Flanking it rise Black Mesa and the three Hopi mesas, immemorial landmarks for all travelers.

The region south of the Grand Cañon is known as the San Francisco Plateau—nearly 3,000 square miles covered with lava flows and dotted with volcanic peaks and craters. In central Arizona once again the high forested mountains sweep in between plateau and desert. Here the broad edge of the Colorado Plateaus narrows to the Mogollon Rim, Mesa or Plateau, as it is variously printed on a map, but better known phonetically as simply the "Muggy Owens." Its southern edge breaks off sharp; and standing on the purple rimrock you look down thousands of feet into the Tonto Basin, full to the brim with jagged crags and peaks.

South to Mexico and east to New Mexico the high plateau resumes. A badland of grotesquely eroded rock surrounded by a desert of cactus, greasewood and mesquite, it was for long the last stronghold of the Chiricahua Apaches, Arizona's most famous cattle range, and a notorious hideout for stagecoach robbers preying on bullion shipments from the Bisbee mines, for Tombstone outlaws, caravans of Mexican smugglers, and bands of cattle rustlers.

Here in the mouth of a cañon in the Chiricahuas is Paradise. No spot could have looked more desolate and forbidding when I rode into it about eight years ago. Somber gray cliffs shut off the wintry afternoon sun. A cold wind swept up from the plain, peppering with sand the thick growth of gnarled blackjacks. But still huddling together, a score of abandoned wooden buildings rattled their loose planks.

Paradise! And on guard at its portals, as if posting its celestial records, was a solitary, narrow-shouldered St. Peter.

He sat in one of the rattletrap buildings with a false front, his high-heeled boots hooked to the rungs of a long-legged stool. In the fading light he was busy sorting old letters in the pigeonholed box on the wall before him. He must have heard me coming up the lonely cañon. Surely he had been aware of a strange visitor poking around the deserted shanties. Yet he did not even look up when my shadow in the doorway darkened the room. Gravely preoccupied, he continued his work like a man on whom a clamoring populace was waiting outside.

The last resident of a deserted ghost town, former justice of the peace and still the postmaster of Paradise.

Actually, except for this aberration, he proved to be a fine old gentleman. Once interrupted he talked generously and well. He spoke of Galeyville, another ghost town of the 1880's three miles off, as though it too were as alive as Paradise. Mildly, reprovingly, as a justice of the peace, he deprecated the occasional disturbances which took place when the boys were in a mood. But like a postmaster, he did not let any unwarranted passion for unseemly law and order obscure his humanitarian view of them as men who also received mail from fellow men.

It was a view of history I have never forgotten. I went

away distinctly comforted by the thought that though Galeyville and Paradise have come to be distorted into the most lurid outlaw camps in the country, and their residents as the most bloodthirsty desperadoes of pulp-paper fiction, they had been in good hands.

Several months later I wrote him a letter, somewhat fearful of the shock it might occasion. The postmaster never received it. He had been transferred to another Paradise to continue serving the former residents of this one in the Chiricahuas.

Northward from Arizona into Utah the Colorado Plateaus extend into the Pahvant, Sanpitch and Tushar plateaus with their tall volcanic peaks. Along the Virgin River in southwestern Utah are three more Painted Deserts exquisite in color, bold and fragile in shape—Cedar Breaks, Bryce and Zion cañons; and eight more great ravines called the Kolob cañons cut into the plateau by minor tributaries. From the mid-point of the eastern boundary of Utah to the mid-point of the southern, the Colorado River slices through the immense Kaiparowits Plateau which comprises nearly one-third of the whole state. Some eighty years ago the Mormon *Deseret News* described it as "one vast contiguity of waste and measuredly valueless, excepting for nomadic purposes, hunting grounds for Indians, and to hold the world together."

No one can add to this description. The immense triangular wedge between the Colorado and San Juan is still one of the most spectacular wastelands in America. Save for a few remote Mormon settlements, only the nomadic Navajos cross it with their flocks of sheep. And only for the last purpose set forth by the *News* can we still imagine its great stone bridges, unequaled throughout the world. Sipapu Bridge, Kachina, Owanchomo and the magnificent Not-se-lid, Rainbow Bridge, as their names testify, are arches that link earth and heaven as well, rainbows of stone.

I remember the first time I crossed it from Moab with a sharp-eyed old-timer named Joe and two cheerful mouse-colored mules drawing our squeaking wagon.

"Looky sharp, boy!" he would proudly observe each morning of the wiry span, making his unfailing joke. "Mules as fart pertest of a mornin' go fart'est by night."

But even this laudable predilection for letting wind did not seem to enable them to progress at all. They only kicked up dust on an invisible treadmill, and we remained forever stationary in the midst of a tranquil immensity that never changed contour. Around us the unending plain of greasy sage spread out under a bright turquoise sky. The same lone Navajo watched us from his flock of sheep. One campfire succeeded another. Weeks passed. Yet even the weather held unchanged and we never moved past the red shapes hovering on the horizon. But occasionally, as if gently to disturb the tranquillity of that vast monotone with a ripple of life, something broke through the crust.

"Looky sharp, boy!" Joe would point out a far, thin plume of smoke. "Injun!"

Days later he would point at it again. "Looky sharp, son! Mormon!"

What distinguished them his taciturnity never revealed. But the latter invariably turned into a solitary Mormon ranchhouse or less often into a tiny straggle of red adobes marked by a precise row of poplars. To the people who came out to meet us Joe always gave the same hearty shout. "Hi, brother!—Hi, sister!" Whether he posed as a visiting Mormon elder or simply imposed on traditional Mormon hospitality I have since wondered. We were always led to the thick-walled feudal fortress flanked by its barns and corrals. Our mules cared for, we lolled in the shade of an enormous cottonwood while a solicitous sunbonneted woman brought us pitchers of cold milk. That evening we were seated at a loaded board with all the family and help. Afterward came the evening reading. The lamplight shone on the Book's pages and on the bearded face bent over it. Around us listening stood Brother Moses—a huge pockmarked Paiute who had greased our wagon hubs, Brother Abraham, Sister Sarah, Sister Mary, and all their children. Simple rough-cut figures with familiar names, they seemed as Biblical as the land itself.

Many times since, a voice from the pulpit has awakened me with allusions to that Promised Land. Moab and Jericho, the muddy red Jordan, take on a comfortable reality. It is all a desert wilderness whose Biblical milk and honey is tinctured with the taste of sage, whose shepherds watch the stars from the folds of Navajo blankets. . . .

Eastward from Utah the sage plains roll into Colorado, lift with a great ground swell and finally break against the wall of mountains. This is mesa and cattle country, one of the last strongholds of the Zane Grey cowboy. Grand Mesa, the largest in Colorado, is its truncated apex. A vast tableland of lava over fifty square miles in area and towering 10,500 feet high.

The country below is more rough and complex. Ringed by the high, precipitous walls of the San Juan and La Plata ranges, the whole southwest corner of Colorado is a region of bold plateaus, jagged-rimmed mesas and steep, rugged cañons jumbled together in almost impassable confusion. Mesa Verde, a green-clad rock 15 miles long and 8 miles wide, is a tangled rough-cut cameo to which few men have penetrated, where straying cattle are quickly lost, and a whole deserted city might exist unknown for seven centuries —as many did.

South and east lies the last eastward extension of the plateaus—the great Jemez Plateau, an immense forested blanket of volcanic tufa, pumice, lava and basalt which drains into the Rio Grande.

But due southward from Mesa Verde you drop suddenly

down upon a vast naked plain marked by the majestic Shiprock and the curve of the muddy San Juan flowing westward into the center of this vast circle which inscribes the plateau region—an area which includes parts of Arizona, Utah, Colorado and New Mexico. The center too of the whole basin of the Colorado; the juncture of the great red river and this its largest tributary, the San Juan. The only place in the country where four states meet, it is known as the Four Corners, the wildest, littlest known and most inaccessible region in America.

You drop down suddenly, as I say, and the years flow backward along that muddy river. The old trading posts at Shiprock, Fruitland and Farmington leap out of empty space, though now surrounded by lush fruit orchards and tidy little farms. We hurried on, Ed Tinker and I, hunting at last the beginning and the end of a boy's almost forgotten dream.

From old Will Evans, Bruce Barnard, Walter Beck, Dick Simpson and Mrs. Palmer, we drove to Townsend's Trading Post, Bloomington, Cañon Blanco and Jim Brinhall's Ranch. Then farther and farther on an old abandoned road.

Have you ever, suddenly and unexpectedly, seen an island rise out of the sea? A place believed long washed away by the surface tides of change? There it was at the end of two sandy wheel tracks through the pelagic plain: the old trading post of Shallow Water, its three-foot thick adobe walls still sound, and lined inside and out with Navajos. The new trader's wife invited us in for a heavy nooday dinner. She didn't remember those two now-legendary Vrain Girls, but of course the miracle happened. Squatting on the floor with two children was a fat Navajo squaw. As a small girl she had been sent to school by Lew and had named her own girl child in English after her.

It is an incident indicative of the enduring unchangeableness of this heart of the American hinterland. The perpetual tides of change alter only the surface. But in its placid depths a single name endures, an adobe wall impassively confronts centuries, and a legend is immortal. And so it is that within this gaping space, untouched by time, exists the unchanged pattern of the oldest civilization in America.

The Pueblo and Cliff Dwellers

I shall never forget the first small group of cliff dwellings I ever saw. Wherever it is, I have never seen it since. The road that led near it twenty-five years ago has been long abandoned; even then it was barely discernible through the scrub cedar. After an hour of rain our Model T Ford bogged, and on foot we struck down into a near-by cañon hunting a dry spot under the lee of the cliffs.

Suddenly, unexpectedly, there it was. It stood across the cañon, set in the sheer face of the opposite wall—a miniature city of soft pink stone. Through the slanting silver rain you saw the dark shaggy cliffs, smelled the rain-washed juniper, heard the rush of water swelling the dry creek bed; and all this was of the familiar sensory world you had always known. But it was that tiny terraced city you felt. An instant, uneasy premonition told you it was deserted. Yet there it stood with a monstrous and enduring placidity; a dead spot in the midst of the vibrant, rushing life around it; an island, fragile but indestructible, in the flow of moving time. It was as if you had stumbled upon something in a new dimension.

Abandon any single house, and year after year its life steadily ebbs away until the very timbers and chimney stones go dead. There is no rejuvenating it by renewed occupancy. So it is to the superlative degree with these abandoned pueblos. Century after century they stand inviolable, untouched, even unseen. Neither grass nor brush nor trees can encroach and reclaim them, so safely isolated are they in a high cranny of the cliffs. Made of stone, they do not decay and crumble in rain and snow. Yucca sandals and pieces of woven cotton cloth embroidered with turkey feathers maintain their texture beneath a few inches of dust on the floor. Even a human body mummifies perfectly in the dry desert air. But little by little the buildings and all they contain go dead—the only true death, without transformation of matter. And finally after centuries they stand as if in a vacuum; like little pink stone castles in a child's glass ball.

These plateaus are full of such strange dead centers—prehistoric cliff and pit dwellings, cave, mesa-top, tufa and enormous ground pueblos.

Mesa Verde, Colorado, is one of the most wonderful Pandora boxes in the world. It contains over 400 mesa-top pueblos and 350 cliff dwellings in its labyrinthian cañons. Escalante, who in 1776 named the mesa, observed ruins in the vicinity but neither climbed the mesa nor explored the ruins. Not until 1888 did two cowboys hunting lost cattle stumble upon the most beautiful of all, Cliff Palace. Really a miniature city 300 feet long, containing 150 rooms, 23 kivas, and both square and circular towers of smooth-hewn stone. Spruce Tree House is nearly as large, with 114 rooms and 8 kivas.

Almost as prolific are the three deep gorges slashed into the red sandstone of the Defiance Plateau in Arizona—Cañon de Chelly, Cañon del Muerto and Monument Cañon. There are over 300 known sites of every kind of prehistoric dwellings, and 134 major ruins. Casa Blanca is the most spectacular. Five stories high, it is so dwarfed by the magnitude of the cliff in which it is set that it looks like a tiny cameo.

Meeting on the west at Chin Lee and spreading out fanwise for a length of fifteen miles, these cañons are among the most stupendous fissures in the whole plateau region. Monument Cañon is named for the obelisks that rise 800 feet high, over half again as high as the Washington

Monument. The sheer walls of Cañon de Chelly are almost as high and scarcely 500 feet apart; and in Cañon del Muerto they are even closer. Nothing disturbs their stark immobility; nothing breaks their profound silence. There are only the massive red walls and between them at the bottom a narrow floor of white quicksand, and at the top a still narrower ribbon of blue sky.

They are still the heart of the Navajo homeland. How at home the people are, as were their ancient predecessors, was shown to me the last time I rode up the cañon with a friend and a visiting ranger. We climbed to Lookout Point. It was late afternoon and small groups of Navajos 350 feet below us in the sandy wash were returning down the cañon to their hogans. But suddenly we were arrested by the sight of a small black speck clambering over the rim of the cliff directly opposite. Slowly, surely, like one of the "human flies" who terrified street crowds a generation ago, he crawled down the smooth perpendicular wall. It seemed unbelievable. Fortunately the ranger had a pair of binoculars, and through these we could see the tiny hand and toe holds dug in the rock, the climber's only aid. He was a Navajo boy who, herding sheep on top of the plateau, was simply climbing down the cliff in preference to walking miles around to get home. We were told later at the trading post that it was not an unusual sight. Once he had been seen climbing down with a lamb flung over his shoulder which had been mauled by coyotes. So too did the ancient cliff builders come and go from their spectacular homes.

In Walnut Cañon, farther south, there are 350 more cliff dwellings, usually in small groups of five or six rooms. On Beaver Creek stands Montezuma's Castle, possibly seven hundred years old, 5 stories high and lodged 75 feet up the side of a perpendicular cliff. A castle for which an Arizona cowpuncher traded a horse and then swapped off for two horses. Down along the Gila and up in Juniper Cove near

Kayenta are subterranean pit houses. There are the great cave pueblos of Betatakin, with 150 rooms, and Kietsiel, with over 250. The ancient Hopi-named mesa-top pueblos of Nalakihu, Teuwalanki and Wupatki of 125 rooms, hardly distinguishable from the weathered rock walls. And finally the great ground pueblos of Pueblo Bonito, Chetro-Ketl, Pueblo del Arroyo, Pecos—immense walled cities with up to 585 rooms.

From Puyé and the Rito de los Frijoles on the eastern slope of the Jemez Range in New Mexico, across Colorado and Arizona to the Lost City of Pueblo Grande of Nevada; from Utah down to Los Muertos built by the ancient Canal Builders along the Gila, these plateaus are sprinkled with prehistoric ruins as remarkable as any in the world.

All the way down into Yucatan uncounted thousands more still await official discovery. Riding horseback through the Sierra Madres of Mexico you come upon them everywhere. From small caves holding old idols secreted from Cortez to great pyramids and whole cities, these vestiges of a vanished civilization cover the continent. They constitute a complete world within a world, a subterranean stratum of life we have never really plumbed.

It is difficult to define the great difference in feeling between the ruins of these Colorado Plateaus and those of the great central plateau of Mexico. The architectural differences are apparent enough. The stupendous pyramids at Cholula and Teotihuacan equal those of Egypt. Monte Alban has produced some of the finest gold- and silverwork ever unearthed. The intricate mosaic walls at Mitla are incomparable. Yet they all are geometically hard and harsh in tone. Their imagery is grotesque and complex of character. Their quality mirrors an inflexible and frozen stylization that borders on a sterilization of feeling; so close, in fact, that it prophesied the approaching death of the civilization itself.

None of our ruins, even those ancient walled city-states

like Pueblo Bonito, compare in size, wealth and variety of detail. But they have a classic simplicity, a fluidity of line and texture that is unequaled. They seem to stem back to an ideology of pure feeling when man maintained a direct relationship with the primal forces of earth and sky through an intuitional apperception of his role in a universe that was an undivided whole. The Aztec was more rationally aware. Conscious of both his power and his transiency, he perpetuated his strength in lofty pyramids to the sun and moon, and officially humbled his pride in public human sacrifices on top of them. The Pueblo was never aware of this separateness from the ultimate source of all power. So the difference between these Pueblo ruins and the ruins of the Aztec empire is not to be accounted for by any expanse of time or distance between them. It is precisely the same ideological difference today between the "primitive" Navajo or Pueblo and the white Euro-American with his equally complex civilization. One is concerned wholly with the internal and eternal; the other with the external over which he has assumed a temporal mastery.

It has always amazed me that we, the most inquisitive people on earth, are so incurious about man himself. Only his habits and habiliments interest us. His living reality we ignore completely. Nothing reflects this attitude more than our prolonged disinterest in these earliest Americans.

The magnificent discovery of Cliff Palace by Dick Wetherill and Charlie Mason roused little interest, if any. It remained for the great Scandinavian archaeologist, Baron Gustaf Nordenskjöld, to explore the ruins. After vain attempts to interest the American government, he was finally obliged to remove his invaluable and irreplaceable collection to a museum at Helsinki, Finland.

The ruins themselves finally protected against "pothunters" and vandalism, we have shown as little interest in their ancient inhabitants. A dreary classification of their skulls and pottery, the establishment of a few dates, and the scientifically false conclusion that most of the pueblo dwellers were of short stature because of their low doorways are the sum of our investigations. All these triumphant and insignificant rationalizations are based on the trite assumption that these people were of Asiatic stock who crossed to this continent by way of Bering Strait and slowly migrated southward.

This is a natural assumption. European emigrants ourselves, we cannot conceive of a race indigenous to America, still psychologically to us the New World. Nor do we like to admit that this vast continent produced an indigenous prehistoric monster or even the "Irish" potato. It was all a great vacuum until vegetable, animal and human migrations conveniently populated it from the Old World in measurable times.

But our primal error in glibly accepting the premise of the Bering migration has been in ignoring completely the belief of the indigenes themselves. Mythology is the only true history of ancient man. And here in the sacred cosmography of Pueblo and Navajo alike, we have a genesis as truly remarkable as that of any race now known. From it emerges the living man, not an ethnological type.

Look at a sacred kiva in any of the oldest ruins or in any of the modern pueblos—and they are identical. It is the universe in miniature. Subterranean, usually round in shape, the walls are the circular horizon. In front of the fire pit is a shallow hole: the sipapu, the entrance to the dark womblike underworld, the place of beginning, where man first came into self-conscious existence. The pine-tree ladder up which man climbed is still the only means of access to the single opening above. Sometimes it rests upon a platform. Thus the whole evolution of man: the floor level representing the second world to which he climbed; the platform, the third; and the roof ladder leading into the outer air, the fourth world, the present world to which he emerged in his successive stages of an evolution long antedating the period of prehistoric monsters and in which the Biblical flood was a comparatively recent incident.

Naked and alone, he watched the shining mountains being made, saw the sun and moon replace their blue luminous light. Birds and beasts took shape. The great forests spread and receded. The grasses grew. Naked and alone man took to subterranean pit houses, to tiny caves like those pitting the cliffs along the Rio Puerco and Rito de los Frijoles. He became gregarious. Great cave pueblos sprang into existence: Betatakin, Kietsiel, and Mummy Cave Ruin, in which has been found a piece of charcoal dating from A.D. 348.

From caves man took to building magnificent pueblos in the faces of cliffs. A horizontal ledge provided support, and the overhanging lip a roof. How beautiful they are really, so high and inaccessible, overlooking valley and desert far below, their smooth walls glowing in the sun.

Growing bolder, the people no longer restricted themselves to the safety of cliff dwellings. They constructed great communal houses on top of the cliffs. All the Hopi mesas were topped with such pueblos. Puyé on the eastern side of the Jemez Plateau in New Mexico, the largest of all, had perhaps 1,600 rooms.

Gradually the people descended from the clifftops, building talus pueblos at the foot of the cliffs against the wall. Of the eleven in the Rito de los Frijoles area, one extended 700 feet along the base of the cliff and was four stories high.

And now at last the great pueblos took shape alone on the plain. One hundred miles south of Mesa Verde, one hundred miles north of Zuñi, one hundred miles east of the Hopi mesas, one hundred miles still from Gallup, the nearest base of supplies—"one hundred miles from anywhere." Here in Chaco Cañon, northwestern New Mexico, they loom out of the dark prehistoric past and stand alone on the sunlit plain. Ancient city-states that rival those of Greece. Walled cities more compact than those of China. Certainly the greatest communal dwellings known.

Pueblo Bonito, Chetro-Ketl and Pueblo del Arroyo, all in Chaco Cañon; Aztec to the northwest with its great 40-foot kiva; Tyuoni to the northeast; Pecos to the south with its vast tenements of 517 and 585 rooms—they are all essentially alike.

A huge round honeycomb covering ten acres like Tyuoni; an immense terraced tenement house containing perhaps 5,000 people, and resembling a great concrete football stadium. From the inner court, 500 feet in diameter, the first terrace rose one story high, the second one story back and two stories high, and so on to a height of six stories with the outside wall solid and unbroken. Or semicircular in form like Pueblo Bonito, its back wall 48 feet high, and containing at least 800 rooms. Or rectangular, like Chetro-Ketl, 440 feet long and 250 feet wide, terraced to a height of five stories. How wonderful they are, really.

Here in this high desert plateau, in the middle of a vast unknown continent, they stood swarming with people. The Church still proscribed as "heretical and unscriptural" the belief that the earth was round. The stones of Westminster Abbey had not been hewn. Marco Polo had not yet journeyed to the court of Kublai Khan.

Then suddenly, about seven hundred years ago, this vast civilization came to a dead end. There was a long drought, we say. The pueblos were deserted and the people vanished, being replaced later by new Indian tribes from the east. So they stand today: the pueblos vast ruins half obliterated by the dust of dry theories; the people theoretical figures classified ethnologically by still later emigrants from another continent. What a monstrous, silly supposition!

Nothing of this civilization has vanished. Neither the

pueblos nor their people. Their history, religion and traditions, still orally perpetuated in the sacred kivas, maintain the record of their unbroken descent.

Oraibi on one of the high Hopi mesas dates back to 1200 when the race supposedly vanished, and it is still occupied today.

Acoma is another. When Fray Marcos returned to Mexico from Zuñi in 1539, he told of having heard of the neighboring "Kingdom of Hacus." Accordingly Coronado, when he reached Zuñi the following year, dispatched Alvarado to explore it. In five days Alvarado reached a village which was on a rock called Acuco.

The rock he saw is one of the most spectacular mesas in the region: a stone island a mile long, rising a sheer 357 feet above the flat empty plain. Its only rival is the equally fabulous rock island of Manhattan. On the latter is built that monstrous, towering cliff city of steel and concrete which is the largest city in the United States. But on the rock of Acuco rests the mesa pueblo of Acoma which rivals Oraibi as the oldest, continuously occupied town in the United States. Its great blocks of terraced houses, each a thousand feet long and three stories high, with its rock reservoirs for holding rain and snow water, were peopled when New York was a desolate wilderness, were old when Columbus first sighted the New World.

When Alvarado first saw it in 1540 its origin was already concealed in a legend reaching back in time to the ancient cliff dwellers, but in space only three miles northward to Katzimo, or the Enchanted Mesa. On top of this rock, 431 feet high and 40 acres in area, the Acomans first dwelt. Laboriously each day they climbed down its single trail to till their small fields below. One harvest time when the whole population was down gathering corn and beans and squash there came a great storm. The rain beat against the cliff, the water rose and ate away the talus at the foot, undermining the huge Ladder Rock which was their only means of ascent. So the people moved to the top of Acuco. Few men since have climbed the Enchanted Mesa even with ropes and extension ladders. But still cut into its sides are the tiny toe holds of the legendary trail, and on its summit and in the talus at its foot are ancient potsherds, stone implements, building stones, shell and turquoise beads—all that remains of the pueblo washed away by a thousand years of storms.

According to a similar tradition, the modern pueblo of Cochiti was founded by the occupants of Tyuoni who followed their leader, Cochiti, in his search for a new site near water. Santa Clara claims Puyé as its ancestral home. Chaco Cañon people moved to Aztec, then wanderers from Mesa Verde. The last inhabitants of Pecos moved to Jemez scarcely a century ago. Even today Walpi on one of the Hopi mesas is gradually being deserted and Polacca is building up at its foot.

Always it has been a fluid population flowing through the centuries from one site to another offering more water, better soil. But always the new pueblo rose in the semblance of the old, whether on top of a mesa or on the plain below. Taos, built in two opposite halves at the foot of its lofty peak, still shows remnants of the old wall that enclosed it. Terraced five stories high and housing nearly eight hundred people, it is one of the most beautiful pieces of architecture in the world. Taos, with Acoma, Zuñi, Laguna and the seven Hopi mesa-top pueblos, and the fifteen ground pueblos along the upper Rio Grande, all perpetuate the ancient pattern.

In a direct line of descent from the unrecorded past, all these vast communal dwelling places so identical in shape and tone are the architectural heritage of America. Nothing else compares with them. Subtly and powerfully through our own few centuries they have made themselves felt. Their terraced shapes repeat themselves in modern Southwest homes.

The towering cliff palaces lining the cañons of Park and Fifth avenues are apartment houses immemorially antedated by those which looked down into the cañons of Mesa Verde. Our new housing projects, complete cities in themselves, look more and more like pueblos—as a Pueblo Indian boy on his way to Bataan pointed out to me in California. Little by little we modernize toward the ancient communal pattern, being dependent with all our technology upon the same ancient verities of earth and sky.

So today the great ruins of the Colorado Plateaus stand as mute commentaries on the transient civilizations and impermanent cities which have grown and ebbed around them. They are not only almost indistinguishable from the shape and color of the plateaus themselves, but the natural expression of the plateaus' dimensional qualities of space and time. The vast space which compressed physical man within great enclosing walls, and which at the same time within the kiva drew spiritual man out of himself toward a living oneness with the universal whole. And that time within which they stand like dead centers impervious to the flow of centuries.

The pueblos are the root, in spirit as in stone, of America itself.

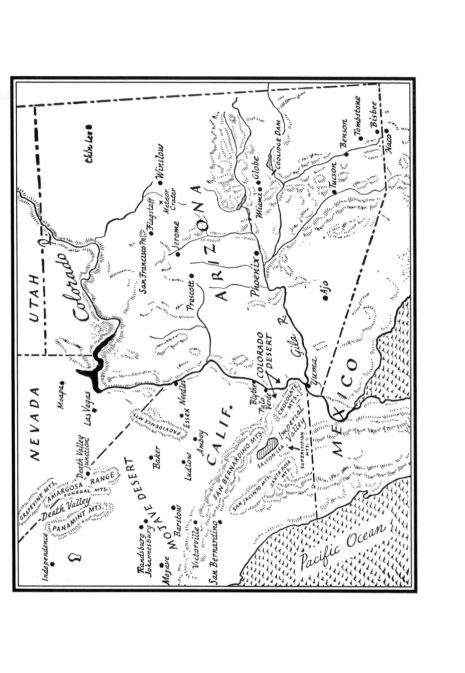

CHAPTER 5

The Desert

THE difference between the Desert and the Painted Desert is one word. No more, no less.

This single adjective applied to the arid plateau region connotes not only all its paint-pot colors, but the singular clarity of the air in which they stand out so fresh and brilliant, and the vibrant quality of the land itself.

Without it there is simply the desert. Lacking clarity of atmosphere, shimmering in the glare of the sun and reflected heat waves, every feature of the land swims in a perpetual violet-tinted haze. Even a discarded whisky flask left lying in the sun gradually takes on the same delicate hue, as bottle collectors have lately found out to their profit. The bare rock mountains are colorless and flat in tone, without sharpness of outline. And lacking the high-keyed vibration, the land lies empty, heavy and dead; like the wrinkled, pitted surface of an old moon.

The desert is the exact antithesis of the high mountain peaks above timberline. Its meaning lies in depth, not height. Time here, too completes its circle. The desert is the below-sea-level floor of the Colorado Pyramid as the peaks are its apex; the end of a world just as high country is its beginning.

There is no better spot to feel the profound difference between these two lower pyramidal levels than on top of the desert range along the Colorado at Goldroad, Arizona. Cross the river at Needles, California, and ascend the steep winding road under a scorching August sun, and you suddenly lift out of one world into another. The air is no longer like the stifling draft from an oven. It is cool and clean and sharp, cutting through your soggy shirt like a knife. In it everything leaps into focus, and in vast perspective there fall away before you the high plateaus extending to the forested mountains.

But stop here a moment if you are traveling in the opposite direction—and you will, not only to deflate your tires, but because it offers one of the most wonderful views I know. Far below, and far as you can see, there stretches an appalling immensity of desolate emptiness cut only by the turgid chocolate river. Nothing holds the eye. Everything dissolves in the haze, even the horizon. It is the desert: ever mysterious, vast, empty and profound; at once terrible and sublime. In one instant, at your first look, you will know it by feeling as well as you ever will. It is impossible to love the desert. But its fatal fascination exerts the most powerful and unbreakable hold of any landscape on earth.

The "Great American Desert" as known by the generation that made the first long trek westward is not to be found upon a map. It included nearly one-sixth of the whole United States and roughly corresponded to the arid region officially defined as that requiring guides to watering places —in effect, practically all of the Colorado River basin. The Great American Desert as we know it today was more clearly defined in 1916 when Congress authorized the expenditure of $10,000 to place signs directing travelers to some 305 desert watering places within an area of 60,000 square miles in California and Arizona.

The desert, then, apart from the arid plateau region and

including part of Nevada as well, is a vast unbroken continuity that includes a half dozen areas distinct in feeling as in physical characteristics.

The great Mojave stretches east from the Sierra Nevadas in California through southern Nevada to Utah, and south to the San Bernardino Mountains where they are cut by San Gorgonio Pass. With an area of more than 25,000 square miles, it is larger than Massachusetts, Rhode Island, Connecticut and New Jersey combined. From 2,000 to 5,000 feet high, it is a vacuous, sandy, wind-blown expanse out of which dust devils rise like sea serpents and race unchecked under the brassy sky.

The molten white heart of this shimmering desolation is Death Valley. The Great Architect could have placed it nowhere else. Its depth balances the comparative height enclosing it, and its walls break the gaping space. Only the Grand Cañon compares with it, yet they are antitheses. The cañon is an intaglio, the valley a cameo.

Walled on the west by the Panamint Mountains, and on the east by the Grapevine, Funeral and Amargosa ranges, all bare rock and rising a sheer two miles in height, Death Valley winds for 150 miles, scarcely 20 miles wile, a shimmering cleft in the earth.

Tomesha, Ground on Fire, early Shoshonean wanderers called it. But seen from the top of the eastern wall—at Dante's View or Daylight Pass—the valley resembles best the white frozen surface of a long serpentine lake. The talcum dust of its smooth white floor, the great alkali flats, deposits of borax, blocks of rock salt and mineral salts washed down from the sierras—all this, played upon by the glaring sun of day or cast with the pale-blue shadows of night, contrives an ever-changing panorama of inconceivable stark brutality and soft incandescent beauty. Standing here, as all tourist folders point out, you look down upon the lowest spot in the United States, 279 feet below sea level, and up at the highest, the snow-covered summit of Mount Whitney nearly 15,000 feet high. If you are without fuel for a campfire, you almost freeze to death at night, and by noon you almost stifle in the heat of its ovenlike side cañons. For years there may not be a drop of rain. Then a sudden downpour will flood the cañons and you will thirst again next day. Confronted by these paradoxes uncounted men have died as if in a trap, and from this valley of death have emerged some of the strangest stories of all America.

I always connect Death Valley with one of my own most weird experiences. At the time I happened to be in charge of the long-distance telephone exchange serving San Bernardino County: the largest county in the United States, over 20,000 square miles, almost all desert. Early one spring I started out with Doc and Carl on a short vacation. My natural excuse was to "inspect" the first telephone installed in Death Valley at Furnace Creek Inn.

With full camping equipment, we took the longest route: from Mojave west to the old mining towns of Randsburg and Johannesburg, over abandoned Wingate Pass, and up into Death Valley from its lower end. This old, dangerous wagon route was supposed to be closed. If one was fool enough to take it going out of the valley, he was requested to notify the office at Furnace Creek before starting. Here he was instructed to telephone back news of his safe arrival at the end of an agreed-upon time. Help was immediately sent out if he was not heard from. But blithely we started out, bucked the sandy road till sunset, and made a dry camp at a poisonous spring near the foot of Wingate Pass.

We ate and rolled in. But tired as we were none of us could sleep. Doc, who had taken me hunting as a boy and whom I knew to be impervious to danger and discomfort, got up first and began dirtying up the mess kit he had just washed. His excuse for restlessness was to try out a new pancake flour. Carl, usually imperturbable, simply walked off and climbed a near-by cliff to watch the moon rise. There was not a soul within a radius of fifty miles. The desert spread out flat and lifeless. The silence was unbroken. Yet something vague, invisible and unfathomable kept intruding upon us all. Suffering under a nameless foreboding and driven by a strange compulsion, we suddenly struck camp at ten o'clock and set off again.

Wingate Pass was a steep horror of sharp volcanic rock in darkness. But pushing and heaving, nearly stripping the gears, we made it. By daylight we were in the dry wash leading into Death Valley. Prying boulders out of the wheel tracks, using chains and old gunny sacks in soft sand, we edged up between the steadily rising walls. Near noon came a new worry. Our water bags were empty and the radiator was steaming. Besides this, we were almost out of gas. By pure luck we drove up to the filling station at Furnace Creek with less than a pint to spare.

Here we got news that confirmed our strange haste. The night before—we were told—an earthquake had destroyed Long Beach where Doc's home was, and half of Los Angeles where Carl, Naomi and my mother lived. Mixed in with this intruded a vision of tangled telephone wires, switchboards lit up like Christmas trees, and my own absence from work during such a crucial emergency.

Thank God for a telephone; it was the only time I have ever appreciated the thing. The new one at Furnace Creek Inn was working and I got through to our chief operator at San Bernardino. She confirmed the news of the quake, but not the amount of damage, for she was unable to establish connections with Long Beach or most of Los Angeles.

It was now four o'clock in the afternoon; and after nearly thirty-six hours of continuous traveling we started back. But this time down the eastern edge of Death Valley to Baker and the main highway leading to San Bernardino.

The office was a madhouse of activity, but all employees had been called in and were working day and night. We continued on to Los Angeles, arriving home late in the afternoon. Fortunately the women were safe, though badly frightened. All the shop windows in the neighborhood were broken, as was every dish in the house. Water and electricity were turned off, the telephone dead. And to add to the confusion a woman passing by at the time had had a miscarriage on the front lawn and was still inside in bed. The family and the doctor were still trying to find out her name. But I would much rather know today by just what means we were disturbed in the middle of the placid Mojave by the earthquake so far away—whether by tremors in the earth or vibrations in the air too faint for conscious perception, but registered by the delicate nervous system.

The whole desert region, particularly the Colorado, is susceptible to constant tremors. Possibly caused by the settling of tremendous deposits of silt at the mouth of the river, they too add to the dimension of depth, a distinctive quality of the desert.

The Colorado Desert is contiguous with the Mojave, and extends southward through California to the peninsula of Lower California. The division is approximated on the map by the San Bernardino, Cottonwood, Chocolate and Chuckawalla mountains. The natural boundary, one might say, is air—the wind driving through San Gorgonio Pass, one of the greatest drafts in the world. From the desert below, the rising hot, dry air tends to create a vacuum into which rushes the cold, moisture-laden air from the Pacific coast. Stopped by the rugged San Bernardino, San Jacinto and Santa Rosa mountains, its only inlet is the pass, and so through this there blows a constant wind. Twenty years ago, before the new highway was built and miles of tamarisk were planted as windbreaks by ranchers, I drove through such terrific sandstorms that the paint was scraped off the car. Such a storm offers a weird experience. The air grows dark as night; one needs lights in the middle of the afternoon. Eddies of white sand ripple across the road like water, and it is necessary to bail out the dust in buckets.

At the bottom of the pass the desert spreads out like a placid sea between the three ranges to the west and the desert mountains to the east. Compared to the high, vast, naked, brutal Mojave, the Colorado seems low, soft, delicate. There is no accounting for this impression. Its peak temperatures exceed those of Death Valley. Its dunes are as desolate. Its rainfall is perhaps lower. Yet its appeal is singularly subtle.

But, like the Mojave, its heart is a sunken depression over 200 miles long and nearly 50 miles wide: the Salton Sink. Nearly 2,200 of its 8,000 square miles is below sea level; its lowest point —248 feet. But if Death Valley resembles a frozen lake, Salton Sink offers a better one: the perfect mirage of a lake lying in the midst of shimmering sand. Strangely enough, it *is* a lake 47 miles long, 17 miles wide and 45 feet deep. Evaporating 7 to 10 feet a year under temperatures that reach 125 degrees, and with an annual rainfall of less than 3 inches, it is kept alive by underground drainage from the surrounding mountains and the overflow drainage from the canals of Imperial Valley to the south.

It is peculiar how the desert continuously offers a resemblance to lakes and seas. Salton Sink gives proof to the paradox. The Gulf of California once extended almost up to San Gorgonio Pass. Gradually it was cut off by the silt of the Colorado River, which built up a great delta across Lower California, forming ancient Lake Cahuilla. The old beach lines still show along the mountainsides on Coral Reef and Travertine Rock, like the rim around a bathtub. It is now believed that this ancient inland lake existed up to fairly late times, and that the present Salton Sink is comparatively recent, receiving its largest fill when the Colorado broke through in 1906.

South and east from the sink to the river lies the **Colorado Desert's** most spectacular feature, the Algodones **Sand Dunes**. Great rippling dunes of clean white sand, it is a turbulent sea of sand. Years ago I crossed it over the old plank road between Holtville and Yuma; sections of it still stick out, upended, in the moving waves on each side of the new highway. Made of railway ties laid flat on top of the ever-shifting dunes, and supporting cross planks just wide enough for one car, it was one of the funniest and most exasperating roads imaginable. Up and down like a toboggan rattled your Model-T Ford. But when on the crest of the dune far ahead you glimpsed another car approaching, there came a race for the nearest offshoot siding. These were locally known as "God damns" and they occasioned as many blows as curses, for stuck between them it was a job for the loser to back up on such a narrow road.

On the Arizona side of the Colorado River and extending south of the Gila to the Mexican border, the desert is known as the Yuma Desert. Here it gives way to the Great Sonoran Desert continuing down the mainland of Mexico to the head of the Gulf. It was from the top of a barren range here that Padre Kino first discovered Lower California to be a peninsula rather than an island.

Across the Colorado River in the upper part of this peninsula the southern extension of the Colorado Desert is known as the Arenoso Desert with its immense Laguna Salada. I remember ruining Carl's old Ford in crossing its high jagged sierras. No short trip could be more diverse in character. From San Diego south to what was then the little fishing village of Ensenada, we were almost continuously in sight of the deep-blue Pacific foaming white against the cliffs. From Ensenada across the peninsula we rose gradually through dense chaparral and the great live oaks to one of the strangest and most charming villages I have ever seen. The Russian village of Guadalupe. Founded in 1904 by a group of fifty Russian families who made few contacts outside, it lay hidden for years from strangers. The thick-walled adobes were thatched with straw; blue-eyed children peeked from the doorways; somber black-bearded Rasputins watched us pass. Past Guadalupe the road was a horror. Hour after hour in low gear we ground upward among low pines to the top of the sierras. Then suddenly the world dropped away at our feet.

We were on the sheer edge of a high, semicircular cliff down which the road wound precariously to the desert below. It glistened in the distance, dry and pale, losing itself in the last bend of the river before it emptied into the gulf. Across this we finally drove to the border at Mexicali.

Below this, the Arenoso Desert, there is still another, the Llanos del Rio Colorado. A desolate coastal plain rippling with dunes along the eastern shore of the gulf, it is the only witness of the great red river's final disappearance in the sea.

All these, then—the Mojave, Colorado, Yuma, Great Sonoran, Arenoso and the Llanos—form one vast unbroken desert from Nevada down through California and Arizona to Mexico and the Gulf.

"Pert near anybody can tell you what it ain't," says the old desert rat, "but there ain't nobody can tell you what it is." A most acute definition. Of only two things I myself am sure. The desert is more than a fact. It is a state of mind. Only to those who have fallen prey to its fatal fascination has it revealed its deepest secrets. They form, as it were, an esoteric brotherhood; and in one place at least they gather for communion:

Where the Elephants Go to Die

Somewhere in the depths of Africa's jungles, so legend says, lies a rock-walled valley where the elephants go to die.

Here, unknown to man, generation after generation of aged pachyderms at last sink heavily to their knees and add their skeletons and ivory tusks to the fabulous wealth of centuries. ...

In the heart of the desert exists this other last rendezvous, almost as strange. Some years ago, the county assessor of that patch of Nevada surrounding it, a distant relative of mine, took me there with him on his annual visit.

It is a small rock-ribbed valley high in a barren desert range. On one side the pale dove-gray desert stretches away into infinity. Far below on the other the muddy Colorado shimmers in the distance, the only break. Isolate and remote, the valley itself is a well of loneliness, a baking oven. Old, rusty mining machinery litters the floor. A few frame shacks huddle against the parched brown walls. There is no sound, no movement save the ceaseless wheeling of a hawk above.

But near sunset the reverberation of a falling rock announces a sudden change. Down from the hidden cañons they come, as out of the remote past. Old stiff-jointed men in blue-denim trousers, naked to the waist and burned black by the sun, with long beards and the matted hair on their chests turned white. In the red twilight the valley comes alive. Knotted old arms chop mesquite. River water is strained and boiled. Salt pork sizzles. Biscuits and flapjacks turn brown. After supper small groups form, playing dominoes, cribbage, poker. But gradually they cluster together in the black, velvety night. Venerable old men with memories like elephants, trumpeting up the past.

In their talk the resplendent days of Cripple Creek, Tonopah and Tombstone flicker in brief phantasmagoria across the cañon walls; the big strikes of Nat Creede, Al Schiefflin and Jim Butler start the pulse anew; the old boom camps of Calico, Rawhide and Bullfrog pop out of the dying embers. This is the legend of America they tell between pipe puffs, and these men know it as no others. In their common odyssey they have followed the lonely burro trails of the Rockies, crossed the plateaus by wagon train, staggered from one desert water hole to another. They have been

hunters, trappers, muleskinners, swampers, ore sorters, laborers; they have driven the spikes of many a rail end. They have known work, danger, hunger, thirst, cold and heat. Only success has eluded them in the guise of fickle Lady Luck. They have outlived disappointment, failure, bitterness—even the world they helped to build.

They receive no mail, no newspapers. Their women are, at best, preserved in a faded, wrinkled letter cached in an old boot. Their sons, if any, have long been absorbed by far cities. They are merely old, homeless prospectors making a few dollars for supplies by working over old dumps for low-grade ore.

But the desert has at last received them and is their common bond. About it they have no illusions; all bear the marks of its claws. But in relating even their narrowest escape, their rheumy blue eyes reflect an amused twinkle. It seems to reassure one that in its profound and desolate emptiness they have found the greatest treasure of all. All else is a mirage. . . .

Vegetably speaking, the desert is the domain of the creosote bush as the plateaus are that of sage. No other resident stands so well the scorching sun, intense heat, alkaline soil and the prolonged lack of water like the 32-month drought of 1909-1912. Other notable desert dwellers are the one-inch galleta grass, the salt bush or desert holly, and the flame-tipped ocotillo. The cacti, almost entirely indigenous to the New World, are bladed like the cholla or seem mere skeletons like the deerhorn cactus. The trees combine all the qualities and appearance of the plants, brush and cacti: the mesquite, paloverde, the stiff, gray skeleton smoke trees that blend in the haze like puffs of smoke, the forests of weird Joshua trees and the giant saguaro cactus. Only the Washingtonia palms, native to the Salton Sink, lend softness to the land and their names to remote oases and watering places like Dos Palmas, Twenty-Nine Palms, Thousand Palms and the now fashionable Palm Cañon.

The paradox is the exotic lushness of their flowers. There are over seven hundred flowering plants in the desert, most of which have pushed their way north from the Sonoran desert of Mexico. The right combination of weather and a bit of rain in the early spring brings them out overnight. For a week at best the desert is transformed into a brilliant garden of purple lupines, pink verbena, white primroses, mariposa lilies, and thousand-hued cactus flowers.

But life here fights hard for survival and the struggle is short. As the plants are all spiny, thorned and covered with a wax veneer, so are the animals constructed the same way to endure sun and conserve water. The lizards, like the Gila monster, covered with scaly armor and able to with-

THE DESERT

stand lack of water and food by drawing on the fat from its tail. Everything is torpid under the sun, but quick as death in movement. All preying on one another, waiting patiently for the first sign of weakness. The little elf owls that top every fence post, gorged on the white-footed mice that dance in the sand in moonlight, themselves watched by buzzards and vultures, dread birds of the sun. The long-billed roadrunner, quick as the rattlesnake or sidewinder it stabs. Ground crickets that migrate into towns like locusts and eat the very clothes in the house. Bloodthirsty horseflies which stab rather than sting, and the soft-footed coyote ever lurking in the distance. Death here is the norm, not life. And the land gives the same impression.

It is the last great battleground of those ancient elements, fire and water. Large alluvial fans gracefully spread out at the mouths of cañons and vast dry lake beds called playas record one effort, as equally vast lava flows, numerous cinder cones and sharp volcanic picachos attest another. And still the terrific heat and the sudden lashing storms continue the everlasting feud. From its results two of the most familiar phenomena of the desert. The peculiar desert mosaic one comes upon so frequently looks as if the desert has been paved with small stones cemented closely together and flattened with a steam roller. Yet it is simply talus rock which has rolled down into a bed of fine alluvium and has been beaten flat by the rain of centuries. Desert varnish, on the other hand, attests the fierce rays of the sun. Everywhere the baked bare rock seems painted by a shiny black oil varnish, in reality the soluble constituents of the rock itself which have been sucked out by the pitiless sun. Too, on a summer day great storm clouds roll in and obscure the sun. Far above you can see the falling rain. But no drop reaches the desert; it is evaporated before it strikes earth, and only the oppressive humidity results.

Yet neither fire nor water has won the ancient struggle.

Only the earth and the air have remained victorious. The air here is a significant adversary. Its dry heat is deadly. One can travel through it for an hour conscious only of its scorching blast, far less uncomfortable than a humid summer afternoon in New York. But at the first hearty swig of water the world begins to spin as if you were suddenly drunk with alcohol. It is the sign that the moisture from your body is being evaporated so quickly that the very cells are being dehydrated. A gallon of water an hour is not an excessive precaution.

At my hand is a current newspaper clipping of a motorist literally baked to death in his automobile while driving a desert road. He was an army captain training here for service in Africa, well supplied with proper manuals and instructions. But he had underestimated the 130-degree heat, and without being aware of it, he was thoroughly dehydrated and lost consciousness. Weighing 180 pounds, he had lost 60 pounds in his four-hour drive.

One of the most horrible sights I ever saw was a group of dead Mexicans brought into Twenty-Nine Palms years ago. Their old car had broken down in the Sheep Hole Mountains. To get in shade, one man, two women and two children had crawled underneath, gradually drinking all the water from the radiator. The sixth, a lad of seventeen, had gone for help. It never arrived. Blinded and delirious before he had walked more than a few miles, he began traveling in circles, and was found dead after the others had been brought in.

Not only does the air kill with its heat—and I remember a truck crashing through the window of the bank in El Centro at 6:00 A.M. when the driver was suddenly overcome. It blinds with its glare, distorts distance with its violet-tinted haze, and plays its most insidious tricks with its mirages. Some of these are vastly amusing. There is a playa in Nevada skirted by the road. Here stands a tiny filling station and red gasoline pump—the only one in sixty miles. There is no mistaking its location. But looking down from the range above, I have seen it washed by rhythmic blue waves; seen it moved miles out into the dry lake bed; seen it reflected, upside down, true in every detail.

Yet at last the land conquers even the air, and through the haze emerge the most fantastic shapes I know, the desert mountains. The Chocolate, Chuckawallas and Cottonwoods, the Shadow Mountains, Turtle, Old Woman, Bullion, New York, Sheep Hole, the Panamint, Grapevine, Funeral, Black and Amargosa, the Santa Rosa and the Cocopahs. Great hairless shapes with dry scaly skins, they look like immense lizards basking in the sun. Like empty, half-deflated balloon-mountains which you could explode with a pinprick. Like heavy molten mountains of lead. Like mountains of the moon. No feature of any landscape on earth exerts their fascinating appeal. Subtly but powerfully they affect even the most unimaginative. Like:

The Man Who Couldn't Turn His
  Back on the Mountains

I knew him just twenty years ago when I was stationed in Imperial Valley, that great green oasis in the Colorado Desert.

The southern transcontinental toll lead had just been built across the desert. Terminating at mammoth switchboards, supplied with intricate phantom circuit equipment, and utilizing repeater stations to boost its fading currents to proper frequency, it was a miracle of science. Over this slender highway rippled all the voices from booming Los Angeles and the West Coast to New York and the great cities of the East. But for nearly a hundred miles between Whitewater and Yuma it was a fragile copper strand stretched across the empty, sun-struck, wind-blasted desert.

A man was needed to patrol it, and only a singular type

would do. Jenkins was finally picked for the job. He was a good lineman, an able mechanic. He was young, healthy, and had just been married. Above all, he was stable and unimaginative, completely extroverted.

The company for its part did its best for him. It boosted his salary. It provided him with a repair truck especially equipped for travel through heavy sand. Then far out in the desert it built him a small bungalow completely furnished with all the latest gadgets. Included was a console Victrola and a library of records. Into this Honeymoon Cottage—rent free!—moved Jenkins and his bride.

It was an ideal job. For weeks at a time Jenkins led the life of Riley. No alarm clock rousted him out of bed to rush to work. Every day was Sunday. He played the phonograph, read heaps of pulp-paper westerns. Occasionally he went duck shooting in a wash not far away, and Mrs. Jenkins learned how to cook out most of their salty flavor. At due intervals he spent a morning in his garden. A fantastic garden: a hundred metal spikes implanted by company engineers to test their various resistances to the corrosive action of alkalines in the soil. Here he would pluck up a spike like a carrot, jot an entry in a notebook, and make a foolish grin. It was all a continual picnic and Jenkins was secretly afraid it might come to a sudden end.

But occasionally, at any moment, the telephone would ring. The transcontinental lead had broken down. Jenkins would jump into his boots and start out in his truck across the desert. Day or night, in the blistering summer heat, in freezing cold, through a sudden lashing rain or the stinging blast of a sandstorm.

Meanwhile testboardmen at their switchboards would be busy calculating the approximate number of miles to the break. So every few miles Jenkins would stop, clamber up a sand-blasted pole and connect with his handset to get their latest calculations, though it was up to him to find the exact location.

On one of these trips I happened to accompany Jenkins. Gradually I began to notice in him a certain idiosyncracy of manner. He would climb up every pole with his spikes, rear back in his safety belt—but always facing the driving sand that cut his cheeks and goggles, clogged his nose and throat, and interfered with the movement of his hands.

"Turn around, man!" I shouted reasonably. "Get your back to the wind and your head down."

But no. Perversely he would shinny up each pole as before, head up against the wind and sand. Even when at last he found the break, and stayed up there an hour to fix it. Preposterous! It not only made his work more difficult, but the delay kept backing up hundreds of transcontinental calls from coast to coast. Also for him personally it was dangerous, and it did not comply with regulation safety measures.

Jenkins finally confessed. "It's those mountains," he said with a foolish, frightened smile. Those wrinkled, dried-up, desert mountains. He just couldn't turn his back on them. They gave a fellow the creeps. It was like they were looking at you all the time, following you wherever you went. Somehow you just had to keep your eye on them all the time. "But some of these fine days, goddam it!" he said with a peculiar exasperation, "I'm just going to walk off from this here truck and go see what's what!"

Jenkins, you understand, was an American.

And these great crinkled shapes floating in the desert haze and splotched with the shadows of clouds in sunshine are the mysterious magnets that have dragged men across a sea and a continent to plumb their empty depth. They are so dead that their deadness has made them seem peculiarly alive in another dimension. Hence they take on a sinister otherworld quality of foreboding fascination. They contain a secret man has always hunted and has never found, but whose search for it has stamped us with its mark forever.

Long before the white man, Indians knew the desert. Our modern highways still follow their old routes. There were three of these ancient highways. One on each side of the Colorado River, running north and south between Nevada and the gulf. Another running east and west through the Mojave from northern Arizona to the Pacific coast. And still another running along the Gila River through southern Arizona, and up through the Colorado Desert and Coachella Valley to San Gorgonio Pass. Along these ancient trade routes were passed all the seashells and parrot feathers to be traded with the pueblos for pottery and turquoise. Great trade routes and highways, they were also the first lines of communication connecting all the tribes and pueblos of the whole Southwest. No better long-distance runners have ever been developed than these Indian messengers, like the Mojaves, inured to running all night, curling up in the scanty shade of a mesquite by day, living on the minimum of water they could carry in a greasy intestine slung over the shoulder. Within a few days of the first arrival of the Spaniards at the head of the gulf, even the Navajos heard the news.

The Paiutes, Mojaves, Yumas and Cocopahs were all desert tribes, almost vanished today. Yet the desert has been a country to pass through rather than to live in, and the heart of its gaping emptiness reflects the absence of man as does no other wilderness. To a stranger the loneliness it engenders is a palpable horror.

I once had a friend, a fellow engineer in the city, who developed pulmonary tuberculosis. The doctors ordered him away from the coast and into the desert. Here, far out on a road from Victorville, he moved into a crude little wooden shanty.

For a year he lived here alone. Every weekend his wife and three small children drove out to see him, carrying books and delicacies. They watched him overcome loneliness, learn how to cook. He grew brown and cheerful, put back on his sparse frame nearly fifty pounds. He was pronounced cured. The joyous day arrived when his family and friends drove out to get him. Before they left the shanty, a sentimental wit, praising the desert's miraculous curative powers, painted on a sign Balzac's definition of the desert:

"The Desert is God Without Humanity."

A little over a year later my friend moved back. He had lost all his regained weight and more. His face was drawn and white, the cheeks stung with telltale red splotches every afternoon. Unable to secure a second leave of absence, he had given up his work for good. All his savings had been exhausted during his previous illness. He had sold his house and furnishings, and his family had moved in with his wife's parents while she hunted work.

It was a grim ride. Husband and wife could no longer keep up their pitiful pretense. Each knew, this time, that he would not return. The desert spread out more tawdry and empty every mile. At the end of the sandy road stood the old shack. The door sagged on its hinges. The window was broken, the floor half covered with sand.

All afternoon they worked. The man with hammer and nails, resting frequently, breathing hard. The woman with broom and mop. Together they hung a bit of bright-colored chintz, the pitiful flag of their last fight. Then dry-eyed, cotton-mouthed, they parted.

The man watched the car dwindle and vanish into sunset. He was still sitting in the doorway when the moon rose, his long thin arms resting on his bony knees. Far off a coyote yapped, the only sound. In the vacuous stillness no light glimmered. It all spread out, interminably, around him—an intolerable emptiness, a tomb already.

He could not sleep. Hours later he walked down to the sign. In the bright moonlight he changed only one word in

Balzac's wording. I saw it after he had died, and it still stands tacked to a leaning post with an arrow pointing to the deserted, crumbling shack. It now reads:

"What is God Without Humanity?"

It might be a commentary, not only on the desert but on the whole vast wilderness of the Colorado. The echo of man's despairing cry against the overwhelming solitude, its haunting, empty space.

And yet it is only a question. The final answer lies in the lives of all the men who have preceded him—in the earliest explorers and Spanish priests who first trod this naked land; in the trappers and prospectors who left their bones to bleach along its trails; in its first settlers who followed uncounted generations of an unknown people antedating even all of these. On them all space and solitude set their mark; they are an intangible dimension of our own lives. They remain, not to be conquered but only to be understood. This is their only secret.

CHAPTER 6

Its Delta

IT HAS often struck me how parallel the course of my own life has run with the Colorado. I could no more write an autobiography without the river than I have been able in this to ignore its fellow traveler. We were both born in the high Colorado Rockies. Progressively in childhood and youth we made our way down the peaks and mountains. Meandering back and forth across mesa and plateau our lives assumed their permanent color, our tempers set. On the desert below we both were harnessed to work for the first time. And the last part of this vast background that I saw, like the river, was its Mexican delta.

Comprising about 2,000 square miles, it extended north and south from the international boundary to the gulf, and east and west from the Sonoran mesa on the mainland to the Cocopah mountains in the peninsula of Lower California.

It was a strange subworld of the Colorado Pyramid.

On the Sonoran side of the river a few ranchers ran cattle and horses on the dry bottomlands. At the upper end, in Mexicali Valley, was the largest single area in the world devoted exclusively to the cultivation of cotton. But below this, from the bajadas of the Cocopah Mountains to the west bank of the Colorado, it was a strange, wild terra incognita to all but a few nomadic groups of Cocopah Indians.

ITS DELTA

It was desert. Mountains. Chaparral. Swamps. Lakes—dry and wet and salt. A crazy quilt of waterways: 600 miles of canals; a dozen channels of the river itself, the Alamo, Abejas, Boat Slough, Rio Paredones, Rio Nuevo, the Pescadero and the Hardy. At low water dry stinking bottomlands, salt-encrusted sinks, alkali flats, tidal flats and geysers of hot mud. At high water a vast bayou.

Flanking the Cocopah Range on the west lay the great Macuata Basin, Laguna Salada, connected with the tidal flats of the lower river at its southern tip. Some 50 miles long, 10 miles wide, and only 12 feet above sea level, it was a vast dry lake bordered by sand dunes. But containing hot springs as queer as the mud volcanoes near Volcano Lake across the range.

Abounding with game, the tule lands about the head of tidewater were full of small wild hogs, descendants of domesticated swine brought in years ago. Wild burros in herds of twenty or more traveled back and forth between water holes. On the highflanking sierras appeared small chamois called "amagoquio" by the Indians. Preying on these were the animals "not unlike the African leopard" which the Patties had noticed a century ago. They might have been the "chimbi," a mountain wildcat with a short tail and beautiful fur, or the lion called "chimbicá," held sacred by the Indians. Larger and more ferocious, it decapitated its victim and after drinking its blood buried the carcass for later eating. Mountain sheep, small deer, fox, rabbits, skunk, muskrats and beaver were common. In the thick chaparral fluttered wild pigeon, quail, doves and occasionally a scrawny turkey. Strange birds with stranger names, "cenzontli" perhaps in Aztec, and "pajaro chollero" in Spanish, the latter suspiciously like an ordinary woodpecker. Buzzards and vultures, hawks, owls and eagles. And above all, the boundless flocks of waterfowl—duck, geese, herons, coots, wrens, sea gulls and pelicans.

It was all one vast contradiction. A jigsaw puzzle confusing as the Mojave Maze. And solved only by the few, shy and decadent Cocopah Indians retreating farther and farther into extinction. One month you might ride on horseback to one of their crude ramadas that a month later you would paddle to in a dugout canoe.

Through this strange wild delta there were no railroads and few roads. The booming Border district of Mexicali was as effectively isolated as if it had been a desert island. It was barred on the north by the United States boundary with its customs duties and all the regulation red tape of immigration and international trade. On the west it was separated from the open Pacific port of Ensenada and the town of Tijuana by a tall range of jagged picachos. Due south the peninsula of Lower California extended in an almost unbroken wilderness. And south and east it was separated from the lower mainland of Mexico by the gulf.

Yet into the Mexican half of the Colorado delta as into the American struggled thousands of new settlers. For the most part these were peons and enterprising young businessmen from Mexico, and Chinese laborers imported by powerful Chinese merchants of San Francisco organized to operate in Mexico under great companies with huge capitalization like the "Compania Chino Mercantil Mexicana." Overland and by water they fought their way northward up from the gulf coasts, and each route bore witness to their tragic struggles. "El Desierto de los Chinos" is still a local name given to a grim stretch of desert where a large company of Chinese laborers perished of thirst. The fate of another group of Mexican peons which attempted the water route is equally tragic.

In November, 1922, the little 36-ton steamer *Topolobampo* chugged out of the beautiful bay of Guaymas on the gulf coast of Sonora, Mexico. She had been named for the bay not far south and was crowded with 125 peons and their families bound for the cottonfields of the Mexicali valley. Little by little the tiny overloaded steamer crept up the coast to the head of the gulf and entered the mouth of the Colorado. At dusk on the night of November 18th the captain ordered the steamer stopped in mid-channel near La Bomba. A small tide was coming in. Two heavy steel hawsers were thrown out to anchor the ship securely on the bar. Tossing lightly on the muddy river, crew and passengers dropped off to sleep.

Near midnight everyone on board was awakened by a terrifying roar. It sounded like a gigantic waterfall booming downriver. Terrified, the passengers rushed to the deck. Clearly in the moonlight they saw traveling upriver with the speed of an express train an immense wall of water nearly fifteen feet high. There was hardly time for a single frightened cry. The wave caught the *Topolobampo* squarely abeam, snapped the hawsers like threads, and rolled the ship over like a log.

Only 39 passengers survived. Days later they were still being dragged out from the mud flats nude, half insane from thirst, blistered by the sun, and raw from predatory swarms of insects. Of the 86 drowned, only 21 bodies were ever recovered. It was one of the worst disasters in the 400-year history of the Colorado, and like Ulloa's first experience it attested the ferocity of the river's phenomenal bores.

But if there were a highway through the delta, Hardy's channel of the Colorado, dangerous as it was, offered the only one available. For several years a courageous and enterprising gentleman named Arnulfo Liera provided the only passage. Under the imposing ownership of his "Compañia de Navegacion del Golfo de California, S.A.," a small steamer ran down the river from El Mayor and La Bomba to Santa Rosalia on the gulf coast of Lower California, and thence across to the port of Guaymas on the mainland of Mexico. Early one August three years after the loss of the

Topolobampo, Señor Liera suggested I make the trip for a pleasant paseo.

Like all things Mexican, the steamer ran by God rather than by schedule. Señor Liera was in no hurry. He waited until he had scraped together a full cargo and a few passengers before dispatching the boat. So every few days I dropped in to see him. Things were progressing nicely. The boat was almost ready to leave. There were already two more passengers sitting on the floor of his office. Two Polish emigrants bound for Mexico City, a man and his wife with innumerable bundles, a case of beer and a caged parrot.

A week later things looked even better. A few green hides to be picked up downriver were being brought by muleback down from the sierras. The two Poles were still waiting, now spread out on the floor of the adobe warehouse on a tattered quilt. A few more days passed. This time Señor Liera was most cheerful. There had been a spell of unseasonable weather. The heat had cracked a seam in the boat, which had to be recalked. Cómo no? Those few hides were still coming down the trail. I saw the two Poles. Unkempt, the man unshaved, they were on their knees sprinkling the bedraggled parrot with water through the battered bars of its cage.

Two days later we were ready to start. Señor Liera was beaming. The office was full of more passengers—peons and their families who were returning to their homeland villages in Mexico. The two Poles were frantic. In broken, almost unintelligible Spanish the man kept asking if there was time for one last drink of cold beer at the corner cantina. "No entiendo, no entiendo," he wailed. The woman began to sob. She wanted to go to the toilet and there wasn't time. There wasn't time.

Next morning at dawn, loaded in an old truck, we started. The voyage had begun.

The Lower River

Unpainted, repellent and inexpressively dirty, the *Rio Colorado* hugged broadside the slimy adobe bank. A small, old steamer of perhaps fifty tons, her 50-foot length was held against the current by a line snagged round a tree. Forward and stern the narrow decks were clear. In the center, divided by a passage wide enough for the open hatchway, were two small deckhouses, each just big enough to hold eight bunks. On the other side of one, protruding toward the rail, was a tiny galley. On the opposite side of the other was the excusado, backed up against the tall smokestack of rusty iron.

Sitting under the tree, I looked for something reassuring about her. There was nothing. I doubted if there was a place on the whole craft to hang my hat. Not only had our own truck been crowded, but another had arrived an hour later to add still more peons to the crowd on the bank. Hour after hour we kept waiting.

It was early morning, but already the heat was so stifling that it seemed to weigh the bright stillness flooding the bare rock hills, the narrow bank and the unbroken solitude of the river's bend. Above us in the sullen cloudless sky hung a hawk motionless as if stuck like a fly to paper. Yards downstream a huddle of wooden shacks and an empty corral were wedged between the water and the upthrust wall of rock. From these filed cargadores, each carrying a heavy green cattle hide bundled with twine. Sweat rolled down their muscular torsos, cutting through the layer of dust raised by their naked feet. Each man trod precariously the narrow plank to the hatch, let fall his hide into the hold, and slouched back down the trail.

Beside me sat the two forlorn Poles dipping water from the river to splash on the parrot. Its bright feathers were dull and bedraggled. It lay on its back with its beak open, one claw raised overhead to grasp the cage. Diseased and panting, it was a pitiful sight until you noticed its eyes. Round, open and clear, they stared back with a satisfied look at once contemptuous and supremely indifferent to the anxious administrations of its benefactors—the look of a spoiled and pampered child.

On the other side lounged a young mestizo named Jimenez. He had made his fortune—nearly two hundred pesos, he confided, and was going back to his village to lord it over his family and home folks. One by one he took off his high buttoned coat which revealed a striped silk shirt, his yellow paper shoes, florid necktie, and stiff straw hat, and then rolled up the legs of his pleated trousers. Lying on his back with one leg upflung, and taking a swig of tequila from time to time like a child at bottle, he looked exactly like the parrot.

Behind us some forty peons squatted with immemorial patience in the blazing sun. The men dark faced and handsome, their strong muscular bodies hidden by shapeless denim. The women meek and submissive as nuns in the folds of their black cotton rebozos. Not a child whimpered.

The smell of hides increased as the hold was gradually filled. Twelve hundred! Señor Liera counted off the last with a broad grin. Immediately the cloud of green riverflies hovering over the landing and the corral swarmed to the boat. Meekly we followed them on board. The two Poles, grabbing up bundles and parrot cage and beer, rushed for the two corner bunks in the rear deckhouse, where at last, completely worn out, they lay throughout the voyage. Jimenez, blind drunk, collapsed beside them. In an instant all bunks filled. The rest of the peons stolidly settled on the stern deck—an immense octopus with dozens of heads, arms and legs.

Suddenly all throats raised a single cry. A huge tree was slowly swirling round the bend upon the boat. At once there answered a calm voice in Spanish. The half dozen cargadores who constituted the crew leapt forward to the rail and with long poles pushed the trunk safely offside. That voice, slow and authoritative, was the only inspiring thing about the *Rio Colorado*. It showed she had a master.

He had been tipped back in his chair against the tiny wheelhouse, asleep in the sun, and now stood imperturbably watching the trunk floating by. He was a full-blooded Cocopah nearly six feet six inches tall and so massively built he appeared fat. The tail of a sweaty blue work shirt hung over his beltless dungarees. His feet were bare, but on his enormous head perched a small white uniform cap. This single and ludicrous badge of authority was unnecessary. Command showed in his face. It was almost black, hairless, with the great nose of initiative and a bony protruding ridge of observation across his lower forehead. Under this his bright black Indian eyes shone hard and steady as a snake's.

A whistle blew. Señor Liera, his truck drivers and a group of horsemen and mule drivers from the corral gathered on the bank. The capitán jerked a bell cord, and the *Rio Colorado* shook with the life of her engines. The plank was drawn on. The line was loosed. Immediately the boat swung offshore; that slow stream masked a current. And when I looked back the clutter of shacks and the group of watchers were no more than a diminutive splotch receding around the bend.

Old, foul with hides and reeking in the blazing sun like a garbage tug, the *Rio Colorado* crawled swiftly downstream. A thin breeze blew in our faces and carried some of the stench behind. I stood on the sloping foredeck, watching the slow red river uncurling like a sluggish serpent ahead. There is always something appallingly monotonous to me in a trip by water; a forlorn sense of sameness and an aloofness from the land that one who prefers burro-back to all other means of travel cannot abide. You will not find it going down the Colorado, perhaps because the river is so inalienably a part of its shores. The chaparral became more and more dense. Shadows of willows darkened the cool glades on each side. Lagoons crept under the overhanging foliage and reappeared between limbs in an aspect of serene and unbroken placidity. Innumerable flocks of cranes and herons floated like spotless white blankets on the muddy brown surface of the river, refusing to be disturbed until we were almost upon them. Then abruptly and silently they rose as if someone had grasped the edge of their feathery blanket and shook it over our heads. And borne forward along that serene and remote watercourse in an attitude of complacent calm, it was as if the doubtful virtues of all our obscure and meaningless lives had been recognized and rewarded with a benign and earthly peace.

The capitán sat backed against the wheelhouse, his bare feet braced against a coil of anchor chain. In this position of easy vigilance he was staring fixedly at the river. From time to time and without turning his head, he murmured a soft command to the boy at the wheel inside. Obedient to his will the *Rio Colorado* veered from midstream into a placid pool gleaming like a plate of bright copper or swung slowly toward the thickets of the opposite shore. Since childhood he had known the river well. He sat there as if it were no more than a brown skein running through his great hands for him to inspect for the slightest flaw. In his hands too he held in bondage our own tangled pasts, the meaningless perplexities of our complex lives.

We were still in this thick twilight jungle when supper was served. Crew and passengers alike stood in line at the galley for a tin plate filled with beans, fried pork, a slice of goat cheese, hard biscuits which exploded with tiny puffs of flour when broken, and a cup of bitter reddish coffee. Darkness fell suddenly. Save for a small riding light on the masthead, the *Rio Colorado* was indistinguishable from the chaparral on each side.

An hour later we stopped and dropped anchor. Immediately silence and heat and mosquitoes settled down upon us. To ward off the latter the peons on the stern deck lighted their charcoal braziers, and in their ruddy glow dropped off to sleep. The bunks of the deckhouses were already filled with the Poles, Jimenez and their lucky companions. On the foredeck the capitán was spread out with the crew around him. Among them on the hard flooring and with my hat for a pillow I lay down. The cry of a night bird rose from the marshes. We slept.

Past midnight a bell sounded. A child awakened by the rattle of the anchor chain set up a plaintive cry. The engines began, and again we slipped downstream in ghostly gray moonlight.

By dawn the peace of night was dispelled. The river had grown to twice its width and reflected a sullen muddy aspect that changed our wholesome mood. Trees and chaparral had vanished. There was nothing but a flat expanse of barren tidal plain stretching toward the gaunt dry hills faintly rising on the horizon. Across this the river wound drunkenly and gracefully in great curves and loops as if it had lost its way. As it straightened, more and more bars appeared. We crept along slowly, like a fat beetle crawling down a dusty road.

Near noon we dropped anchor again. The day dragged by and still the *Rio Colorado* did not move from her mooring fifty feet from shore. For what were we waiting? Only the capitán knew. But impassive and unmoving, perhaps asleep, he lay there as if oblivious of heat, sun, flies and our own impatience. The river, slow, indifferent and majestic, crept by us as though we, heirs to power over all the living earth and its stubborn elements, were no more than an insignificant and unmoving speck upon its muddy surface.

It did not know us; and we, with resentful, sun-glazed eyes, stared back upon the river.

For three days and four nights we lay as if frozen in time and abandoned to the capricious stubbornness of that massive Indian. Perhaps fifty people crowded on a flat board surface 50 feet long and 20 feet wide, anchored in the middle of a muddy river flowing through a flat sandy plain, and exposed to the glaring desert sun of August. Psychologists assert that a dream exists no longer than a second in a sleeper's mind, and only the most foolish of us deny that a man may live longer in the stress of a single hour than during a year of easy life. So those interminable days on the *Rio Colorado* comprise a period which remains timeless and immeasurable, like a fragment of life curiously detached from the moving stream of years.

Lackadaisically each morning at sunrise the mongrel crew scrubbed the deck and wiped off the rail. The rest of the day they lay in every torpid shadow on the boat. The two Poles never ventured out of their bunks, exemplifying by their patience a belief in the invincibility of time to remedy all ills. Jimenez was a bore. Running out of tequila he would lie awake at night, then reach out and jerk a feather from their parrot. At its squawk the Poles would awake, curse and plead, then finally pass over to him a bottle of their treasured warm beer. The peons on the stern crouched patiently under their canvas flaps, their naked feet sluffing to and from the water cask wearing a trail into the planks. Courteous and soft spoken, they never passed without murmuring "Permiso, Yanqui" or a slow "Con su permiso, señor."

The fried pork gave out. Thereafter we ate beans, cheese and biscuits three times daily. The uncovered metal water casks developed a thick green scum from beneath which we dipped stale and tepid drinking water. The toilet became a horror against whose rusty back we burned our own whenever the engines were running. The hatch was opened to let in air to the hides, and the sun circling overhead drew from the hold a stench that made us dizzy by midday. I would have given anything for a bar of soap and a toothbrush, but my duffelbag had been thrown in the hold and could not be found.

Not a soul on board spoke English, and this with my incomplete Spanish set me apart in the unenviable position of a solitary white Yanqui in the midst of an alien race. The two Poles were curiously self-isolated with their diseased and pampered parrot. Jimenez with his pathetic illusion of superiority derived from the money hidden in his waist was separated from the ill-kempt crew of Mexicans; and with the crew from the horde of peons by a greater gulf—the impassable breach between the mestizo and the Indian. Between the capitán and crew and passengers alike lay an intangible barrier composed of authority, race and his own temperament. He was a man who would have been alone in any crowd. So that jammed together as we were, and weighted down by the same miseries, we were at once a world isolated in space and time, and a dozen worlds.

Feliz alone expressed our common humanity. He was a little shirtless fellow about eight years old, the eldest of four children of one of the señoras nesting on the stern deck, and was working his passage. Next to the capitán he assumed all the responsibilities of the *Rio Colorado*. If he had not, they would have been thrust upon him. For two hours, three times a day, he washed tin plates and scrubbed greasy pots in a wooden tub in the narrow passage outside the galley. The filthy ill-tempered cook compelled him to haul his own water buckets up the side of the boat, an almost impossible feat. Once at the lurch of the steamer he almost went overboard, a diminutive weight hanging grimly to the end of his line. Luckily the capitán was passing by and caught him by the heels. Thereafter Feliz substituted a tin pail for the heavy bucket. Between times he did chores for the crew, waited on the capitán, carried water to the peons and coffee to Jimenez. Answering to the call of "muchacho! —muchacho!" he was kept on the jump day and night. The only time I ever caught him loafing he was munching on one of the hard floury biscuits with the rapidity and ferocity of a squirrel. His precocious little face, always unwashed and sharpened with hunger, already had assumed the solemn gravity of a peon. Yet throughout he maintained an unsmiling good nature. Invulnerable to jests and curses, denying pity for his lot with a subtle self-reliance, he always carried about him a staunch reassurance that he found the world agreeable enough without rest or play.

Restless, anxious and bored, suffering heat, sun, mosquitoes and the stench of hides, we continued to watch with dull glazed eyes the chocolate river oozing past. Only the nights seemed real. Under the moon the river took on an aspect of serenity and mellowed age, and its broad gentle windings seemed like a road that had been long abandoned. Far out midstream a fish rose leaping, and its splash might have been a tuft of dust. Forgetting each day as we had forgotten the others, yet unreasonably hopeful of the morrow, we lay uncovered on the hard decks. And steadily, hour after hour, a boy sitting on a water cask kept taking soundings of the river. We heard the splash of lead, its bump against the rail, and in a moment his low soft voice rising into the night. "Cinco brazas, capitán."—"Cuatro y media." —"Cuatro!"—"Seis, capitán."

That massive Indian was still awake and waiting. One night it came. The river rose rapidly. The *Rio Colorado* jerked at her anchor chain. The capitán stirred for the first time. The anchor was brought up and the engines started, though we did not move. Greasy lanterns were lit, and in their dim flicker the crew lounged restlessly, rolling cigarettes. Far downstream sounded a low resounding boom. Swiftly it advanced upstream upon us, clearly visible in the moonlight: a wall of water some four feet high sweeping round the bend. The *Rio Colorado,* unfettered and with engines running, met it squarely. She went nose down until her decks were washed and came up with a dizzy roll streaming torrents from every passageway.

The capitán seemed pleased at her ducking. Wet to the knees, he stood listening to the receding roar as if oblivious to the commotion on board. It was punctuated by the blasts of a raucous siren which turned out to be the squawking parrot. People poured from the flooded deckhouse. Foremost came the two Poles carrying the parrot. Its cage had been sitting on the floor beside their bunk; the flood had dunked it thoroughly; and now, indignant at having been doused instead of sprinkled, the bedraggled lump of feathers was screaming with rage. Jimenez clung to his stiff straw hat, his money belt and yellow paper shoes—all that would differentiate him as a pompous man of fortune among his simple villagers at home. Feliz, eyes scared to twice their size, still munched absent-mindedly at a biscuit as though his unconscious hunger would accept nothing short of drowning as a surcease from its task. Only the peons made no sound. With peasant stolidity they endured water as they had endured sun, secure in the unrelenting justice of their fate. But we, frightened into an abject mass clinging to doorposts, deck rails and bunks, shivered in abysmal misery wringing out our clothes, and waited for the sun.

Impassively the capitán ordered the crew to drop sounding line and anchor again. The engines were stilled. Again we lay waiting in the river.

What we had experienced was one of the tide bores of the Colorado. And what we had escaped, due to the capitán's caution in waiting far up the channel until its full force was spent, was the fate of the *Topolobampo* three years before.

A bore is simply a tidal wave which rushes up between the narrow converging shores of a river in a high and advancing wall of water opposing in turbulent conflict the current of the river itself. Such bores are encountered in but few river mouths in the world. The Hangchow Bore at the mouth of the Tsientang in China, and the Shat El Arab where the Tigris and Euphrates unite to flow into the Gulf of Persia are notable examples. Bore waves rising from 6 to 9 feet high are known at the mouths of the Severn in England, the Seine in France, and the Hugli in India; waves 15 feet high have been known in the Amazon. Of all these the bores of the Colorado are perhaps the most phenomenal. For here the tidal variation of the Gulf of California ranges from 22 feet to a high of 32 feet, rhythmically generating tremendous waves that sweep up the river for 37 miles.

Opposed to this is the phenomenal force of the Colorado itself. Perhaps the heaviest silt-laden river in the world, it sweeps down into the gulf at the rate of 200,000 cubic feet per second.

Always strange, terrifying and dangerous, the tremendous conflicts of these opposing forces vary in season with the flood, high- and low-water discharges of the river; with the monthly phases of the moon which control the ascending tides of the gulf; and with the distance upriver from the mouth. Yet from Ulloa's first experience in 1539 to the disaster of the *Topolobampo* in 1922, every boatman who has witnessed the bores has attested their ferocity.

We had been lucky; it was August and the river was at its summer low; also the tide of the full moon was past its peak. But our huge, taciturn capitán was taking no chances. With an eye on the waning moon and an ear cocked to his boy's cry of diminishing soundings, he lounged patiently on the still-anchored *Rio Colorado*.

* * *

The river fell eight feet, ten. We awoke one morning in what seemed a different channel. The flat edge of the plain was now a steep, muddy bank rising above our heads. Downstream the river parted to each side of a sand bar already drying in the sun.

But the capitán could not be hurried; it was almost noon before we started. The sound of the engines warming up and the boat's vibration shook us alive. With something of a shock we remembered we were on a boat and had a destination. Imperceptibly, strangely enough, the *Rio Colorado* began moving past the bar toward La Bomba.

The histories of most great rivers are written in the cities at their mouths. Paris, London, Buenos Aires, New Orleans—they are all no more than what the Seine, the Thames, La Plata and the Mississippi have made them. But it is a singular characteristic of the Colorado that it has no city at its mouth. Like all the others it was the first known highway into a new and unknown wilderness; for three centuries it was the clearest trail into the deep hinterland of our America; and its mouth was the only portal. Strategically located at the head of an immense gulf, it could have commanded the dominance of a peninsular province and its motherland shore, the trade of two countries and one of the greatest fishing grounds in the world. Yet only La Bomba remains as a monument to what might have been the first and greatest city in the New World. It lacked only the one thing that would have made its development possible, a rich, accessible hinterland instead of a terrifying desert.

A dreary, sun-struck huddle of deserted shacks, a crumbling loading pier—no more than this, it slips past without a hail. It is not even a remote little fishing village like San Luis Gonzaga, a watering place like El Doctoro, a boat landing like El Mayor. It is simply an Almayer's Folly that today has no name upon a map. Here for four hundred years or more Cocopah dugout canoes, Spanish galleons, Hardy's *Bruja,* all the early American steamboats and Mexican fishing boats have stopped. And so today it persistently exists in the memory of every man who has ever run the lower river, one of the most familiar names throughout the gulf. And yet as if by some strange miscalculation of fate, it remains a site that has missed a resplendent destiny by a hair.

Slowly we nosed past it into the broad highway of the sea. Almost two kilometros wide, the river kept spreading out. What was river, gulf, tidal plain, mud flats and fields of alkali and salt no one could distinguish in the blazing sea of heat waves shimmering in the sun. For here at its mouth the Colorado, gorged with silt, is as much land as water. Looking down over the rail we seemed to be afloat in what resembled a vast settling basin or a vat of ore concentrates red as burnished copper.

Such areas of extreme sedimentation have given rise to many strange phenomena. On June 16, 1819, at Cutch, India, one of the strangest occurred in the Indus delta. Without warning 2,000 square miles of land suddenly sank beneath the sea. The shock of this tremendous displacement spread throughout an area within a radius of 1,000 miles. The town of Bhooj collapsed in ruins. At Ahmedabad, 200 miles away, the famous mosque built 450 years before was shaken to pieces. To the northwest the Denodur Volcano erupted. As the shock subsided, the sea rushed in, inundating miles of land. Then suddenly and mysteriously the "Ullah Bund," or the "Mountain of God," lifted bodily out of the level plain.

Little wonder that men fell on their knees, though powerless in prayer to stop, control or even understand this manifestation of a mysterious power that in an instant could wipe out cities, alter history and forever change the face of the land.

And yet what had taken place was simple enough even

to the most ignorant. Like man, the earth but trembled on the scales of God. When one area went down beneath the sea another rose to maintain the balance. It is the perpetual, recurring history of islands and mountain peaks, of continents themselves. "Isostatic equilibrium," Grandfather's old phrase; it holds as true at the mouth of the Colorado as it did on a high peak above timberline at its source.

From this has evolved the Subsidence Theory, which holds simply that the sinking of land at the mouths of rivers is due to the enormous weight of the deltas breaking through the crust of the earth.

According to this, the great earthquake of Lisbon in 1775, at Cutch in 1819 and at Cachar, India, on the Ganges, and at Charleston in 1886 were all caused by the subsidence of deltal areas. Even the San Francisco earthquake of 1906 was aided by the subsidence of the heavy silt deposits of the Columbia River brought southward by the drift of coast currents in the Pacific.

Consider the Colorado delta, about which J. O. Turle, a geographical engineer, collected in 1928 some amazing facts for his study. It is one of the greatest accumulations of silt in the world. The whole area south of Imperial and Coachella valleys to the gulf is but a vast delta bar built up by the Colorado's deposition of silt. The river has gouged out of the Grand Cañon alone 350 cubic miles of rock, which has been deposited at its mouth. And still every day it carries through the Grand Cañon an average of one million tons of sand, equivalent to 80,000 railroad carloads, to pour on top of them.

A subsidence of less than 50 feet of this enormous weight would cause the gulf to rush up into Imperial Valley and cover the whole Colorado Desert—a catastrophe beside which the flooding of the river itself would seem trivial. What are the prospects?

It is true that areas of sedimentation occur near the

mouths of all large rivers. Yet in almost all cases tidewaters sweeping along the seacoast carry away the silt, distributing it equally over the ocean floor. The Gulf of California, narrow and landlocked, offers no such release. Furthermore, the rate of subsidence of these areas at the mouths of rivers is nearly proportional to the ratio of their drainage area to the area of their deposition. This average ratio is 16 to 1. That of the Colorado—draining 246,000 square miles and depositing upon less than 1,000 square miles—is 250 to 1, the highest in the world. Also in the former average, the usual rate of subsidence is from 2 to 4 feet each hundred years, whereas at the mouth of the Colorado it is 44 feet.

The proof of a pudding is in the eating. What about the theory? The answer, according to Mr. Turle, was obvious to geological engineers and laymen alike. Hardly a month went by without a minor earthquake in Imperial Valley rattling windows and cracking another cornice of the Barbara Worth Hotel. A few years later he could have offered another proof with the serious earthquake of Long Beach and Los Angeles. A map of the great rift faults of Southern California, the San Andreas rift fault north of Yuma, the Salton Sink blocks, and others converging at San Gorgonio Pass show that all were ready to go down under weight. Unless—as had been pointed out—an immense dam of some kind was built to stop the deposition of silt at the mouth of the Colorado.

But to us on board the *Rio Colorado* that fantastic project seemed as hazy as the far-off volcanic sierras floating on the horizon. We were more interested in the water content of this strange area than in its discoloring silt. It was still as thick and dark as chocolate as far as we could see. But somewhere in this vast dark upper end of the gulf there was a strange white circle of clear, fresh water bubbling up through silt and salt water. In it, completely marooned, so to speak, lived a school of strange fish found nowhere else in either river or gulf. Now these fish, breeding at such a rate as soon to overpopulate their tiny pool, had developed a peculiar custom. The females were ferocious, bullheaded and much stronger than the males. So each year, as the spawning season passed, these amazons would draw up in battle array and rush upon the males. Wriggling, butting with their bullheads like great stags and slashing with their teeth, they would force the unneeded males out of the circle where they would be choked in silt, drowned in salt water or burned to death in the lagoons on either side.

For also in the mouth of the river were two lagoons of reddish water of such a caustic quality and so malignant that a mere drop on any part of a human body creates blisters and burns into the bone like acid if not wiped off. This was probably caused by a bituminous mineral on the beds of the lagoons, which has often eaten up the anchors of fishing boats by its corrosive action.

It was the ingeniero of the *Rio Colorado* who vouched for both these tales. He had just come up from his boiling engines down below, and stood reeking with sweat and grime. He was a talkative fellow. Born in a remote fishing village along the gulf, he still talked of fish. Besides his engines they were all he knew or cared about. All the grease and oil in which he seemed invariably clothed, as if in the raiment of his new calling, never obscured the imagined smell of fish about him. Even his long drooping mustaches seemed perpetually awiggle as he talked.

"This gulf, señor, is a garden of fish. They sprout here faster and thicker than corn on the hillsides." And his greasy arms, sweeping in generous, careless arcs over the water, seemed to mark off those pastures of the sea as a ranchero on a hilltop marks off his crops and herds. He was particularly enthusiastic over the totoaba which he called the "crow." Of steel-blue color, and weighing up to 165 pounds, the totoaba breed in greatest numbers in the mouth of the

Colorado which they enter for alimentation in the shallower water. Fishing for them is not much sport; once hooked they are pulled in like a cow on a halter. Only their size impresses a visiting Yanqui, and the poor fisherfolk are always glad to get them for the "buche," or bladder, which can be sold. There was also el lobo marino—the sea wolf; the atun; the delicious dorado, a gold finch; the palometa, a perch with four blue lines running across its back; the huachinango; the pez-gallo, or flying fish; the bagre tarpon with two white mustaches hanging from its pouting lower lip; the mammals cochinillo and tintorera, which change colors when dying; the valuable shark whose fins were sold to the Chinese in Mexicali; together with tortoises, sea hogs and water eels.

The ingeniero's voice ran on. Then after a lurch to the rail to point out a flash in a trough, he recalled sadly, "But I, señor, I am but an engineer," and clattered back down the steps into the hold.

The *Rio Colorado* kept nosing into the gulf, plunging uneasily in choppy waves that gradually cleared of silt. Suddenly we noticed that they were blue, bright blue. We were out of the river, past the islands at its mouth, and abroad in the open sea.

Slowly our little vapor crawled down the desolate coast of the peninsula. Caught between the great blue sea and the jagged, black, volcanic sierras of the shore, she seemed to have shrunk perceptibly. But only to increase the stature of her massive Indian master. The capitán had come to be my best friend on board. Day and night he lay beside me in his habitual posture of watchful ease. He was as familiar with the gulf as with the river, and his eyes did not miss a point or cape. And those sharp, barren outcroppings of rock, imprinted as indelibly upon his mind as upon the horizon, seemed like symbols of his life's corners, like cryptic sign posts of his domain. The *Rio Colorado* was his life and its horizons his chosen world. Sure of it, certain of himself, he carried an air of unmarred self-content.

Toward evening a squall blew up. The Sea of Cortez has always been known for its rough passage. Luckily we were too early for El Cordonazo de Francisco, that hurricane which invariably occurs on the saint's day of San Francisco and is dreaded throughout the gulf. Nevertheless, we crept into a small bay. Instantly the boat ceased pitching and lay so quiet in the darkness I must have fallen asleep. Near midnight I was roused by the capitán's hand laid lightly on my shoulder. For what I never knew. He did not speak or turn around, but remained staring fixedly across the still water.

The moon was up, and over that remote and placid pool it gleamed silver as the shield of Cortez himself. In it, in profound silence, slept the *Rio Colorado* enclosed by a great half circle of somber cliffs. There was no movement of the boat, but we could hear the resonant thunder of high waves against the rocks and with it the low undertone of wind rushing by the entrance of the bay. I have always remembered it and that touch upon my shoulder. But what could you say to a man like that?

Next morning we put out again, stung by a fine spray blowing across decks. We were in a storm. The capitán said so. To a landsman a storm at sea carries the reasonable expectation of something awe-inspiring and magnificent. Yet there was nothing to see. The sea was merely choppy, the waves breaking into tiny whorls of white spume like hillocks of prairie grass stirred by the wind.

Among them the *Rio Colorado* staggered like a locoed cow. Between jerky pitchings over their crests she lay wallowing from side to side in the troughs. Only occasionally did she meet a wave head-to with any drive at all, and then the surge, just deep enough to be disagreeable, drained down the dirty decks. We might have been in a rowboat for all our feeling toward that 50-foot steamer. For not once had the *Rio Colorado* proved to us her superiority over her medium. Frightened equally by river and sea she seemed a trespasser fleeing for a port to hide in. True, she had kept afloat, but any log might have done the same, and we had long given up the feeling of going anywhere. Imprisoned on that miserable craft we could only adjust our lives to hopeless bondage until our time was up.

The first of several huge swells caught us at an angle. All morning we had been keeping our internal equilibrium only by an effort of our wills. That swell caught us off guard. One by one we gave up the ghost and lay weak and sweaty on the planks. Four of the crew crawled into the lifeboat, where they collapsed retching over its sides upon the deck. Even the capitán retreated to sit behind the boy at the wheel, refusing even coffee. Certainly the wet foredeck was no place to spend that night. I went to the stern deck. One of the women had built a tiny fire in a charcoal brazier, and its flicker revealed a shapeless, intertwined mass of naked, shivering children, women and men spread out upon the flooring and convulsing at every lurch. The terrible stench drove me to the deckhouse. It was worse.

Things in that 12-foot cube were indescribable and unbearable. In bunks scarcely wide enough for one person, two and even three lay together like limp bundles of rags. The floor was a solid mass of peons who had crawled in to escape the flying spray. Upon these, those in the upper bunks vomited at will.

The night dragged by. Another day passed. The pitching of the boat had ceased but still we could not eat. The huge pot of beans, untouched for two days, stood in the passageway crawling with maggots. The water casks were filthy with algae as if the water had been dipped up from a marsh. Tin cups and plates littered the *Rio Colorado* from bow to stern. Her decks were slimy and encrusted with salt.

We and she needed a hose more than anything else, but weak and dizzy, we shivered the instant the sun went down. Yet the engines going at top speed had so heated the rusty side of the smokestack forming the back wall of the excusado that it could not even be approached. The *Rio Colorado* was not a squeamish place. We were all reduced to our common humanity at last, an integrated whole.

In this immense and haunting delirium I saw Jimenez staggering out on deck. He had mashed his stiff straw hat. His silk shirt, slept in for a week, was stripped of buttons and hung tail out over his pleated trousers. They were splotched with vomit. Then I saw his sick, anguished face. It had turned a bright yellow, even to the whites of his eyes: yellow jaundice. And seeing him I knew I was looking at a mirror.

He staggered to the rail and retched. Instantly from the surface of the water heaved a dark slimy body that resolved for an instant in a row of gleaming teeth. Sharks. All day they followed us down the coast, a row of thin knife-edge fins cutting the blue.

But we were moving. Out of the gray watery plain rose La Isla Angel de la Guarda. Between it and the Bahia de Los Angeles we crept down through the channel of Sal-si-puede—Get-out-if-you-can—to the island of San Lorenzo. Eastward loomed the island of San Esteban and still farther, as if it were the mainland, rose the brown rock walls of Tiburon, the last home of cannibals.

Still we kept moving under a bright and merciless sun, kept moving across a sea that was as hard and smooth as a plate of stainless steel. But drawing closer to shore to see the jutting bare brown cliffs. Punta San Juan Bautista, Punta Trinidad, Punta Baja . . .

"Cabo Virgenes, señor. We are here."

It was the capitán who had spoken. He had put on another shirt, tail in for once, an enormous pair of bright yellow shoes with glass buttons and "bulldog" toes, and his ludicrous marine cap.

The engines slowed. High upon the scorched hills stood a huddle of board shacks. A little farther and we saw a long breakwater, a rattletrap jetty and a single palm. The engines stopped and immediately a torpid silence embalmed the boat. "Santa Rosalia, amigo. . . . Señor Yanqui, we are here!" The capitán nudged me again. I thought I had best be getting with him down into a boat. Perhaps there would be a comfortable toilet, a bar of soap and a toothbrush on shore.

Two evenings later we left Santa Rosalia. It was still merely a French concession of the Boleo Mining Company: a great smelter and a group of aloof living quarters on top of the hill, and below it a tiny plaza surrounded by the squalid adobes of native Mexican workers, formerly Yaqui slaves brought across the gulf from Sonora.

The *Rio Colorado* had been cleaned up considerably and the other passengers, with fresh water and food, looked better. Jimenez, skin and eyes more sickly yellow than ever, had put on a new silk shirt without removing its price and size tags. Little Feliz was contentedly munching on a banana. The two Poles at last had ventured from their bunks, and even the parrot had perked up. They were all gathered on the foredeck about a new passenger sitting—of all things—in a deck chair. She was a Frenchwoman about sixty years old who had come from Marseilles in time for the birth of her grandson. The father, she said, was employed by the Boleo Company and still had two years to serve.

The *Rio Colorado* slid quietly eastward across the glazed, deep-blue expanse of gulf. Dark came on. All night long we could see the vivid flashes of sheet lightning over the Sonora highland nearly a hundred kilometros away. And then, awakening suddenly to what it meant, we saw a faint dark blur of hills. Guaymas!

In the ghostly darkness of early morning we entered

the outer harbor. Not only to me and the capitán is that landlocked harbor the most beautiful in the world. We seemed lost in a maze of hills; then suddenly, magically, they fell away. We were sliding across a buckler of smooth steel studded with the bolts of tiny islands. We passed a small packet standing under a load of sail. We crawled, black and brutish and still stinking, past a pleasure schooner spotless white, with a sheer as trim and lovely as a lady's waist. That was a ship! Yet what could she do an hour up the Colorado? Somehow for the first time the *Rio Colorado* seemed other than what she was, humble and courageous, forever familiar and ineffably dear.

Offside a voice eerie in the early morning cried out and was answered by our capitán. A rampart of hills rose up and blocked our passage. The engines stopped. Waiting for daylight we washed from buckets hauled up the side. The capitán put on his shoes again and got out his papers. Jimenez nervously wrapped round his throat a new silk scarf.

Gradually the darkness thinned to a sullen gray and suddenly it was light. Those hills were green. Vivid green! Imagine green hills! The sierras of Mexico. And below them the tall towers of a cathedral, a row of palms, a long embarcadero. Home! For still and always Mexico remains the spiritual mother of that vast wild province of the north, the wilderness basin of the Colorado. It seemed suddenly, for no other reason one could rationalize, that after four hundred years we had at last returned home. And with the first shaft of the sun clearing the sierras and striking into the stainless bright blue of the bay, a boat put out to us from shore.

PART TWO
Its People

CHAPTER I

The Conquerors

IT IS WRITTEN in words like these that when Cortez planted his first Spanish cross on the shores of Mexico the fate of the New World was determined. Other words and cryptic signs had prophesied it long before. The return of the white and bearded, maizelike but human form of the Plumed Serpent, Quetzalcoatl, was anticipated at the end of a cycle which we know now corresponded to the last of the Aztec Empire. Comets and earthquakes heralded his coming. Supernatural voices warned Moctezuma of his own and his country's death. Strange boats like birds were envisioned a year before the coming of Cortez. But it was the conqueror who came, and with the miracle of his conquest fulfilled the prophecies. And an empire, a civilization, the whole continent was thenceforth dedicated to a new destiny.

The white newcomers themselves brought along fable and myth to influence their greed. The story of seven rich cities in an unknown land called Cibola was told long before Columbus rediscovered America. Somewhere in the Gran Quivera its emperor slept under a tree hung with golden bells. Elsewhere reigned El Dorado, the Gilded Man, whose freshly oiled body each morning was sprinkled with gold dust and washed off each evening in a lake whose shores gradually became sands of gold. And so El Dorado became at once a man, a lake, a city, a land—a dream to suit every dreamer. There existed also an undiscovered island called California populated only by women and governed by a fabulous virgin named Queen Califa, where gold and pearls were to be found in great abundance.

A book carried in the baggage of Cortez's soldiers authenticated the latter. It was an old Spanish romance called *Las Sergas de Esplandian* written about the end of the fifteenth century by Garcia Ordonez de Montalva, and supposedly the first book sent to the fire when the curate and the barber burned the library of Don Quixote. Conjecture has it that the author obtained the name of the island California from the famous French epic poem *La Chanson de Roland* of the eleventh century, for in it likewise appeared allusions to a marvelous land called "Californe."

Cortez could not question its existence after seeing Tenochtitlan. His conquest of the mainland secure, he sent forth under Don Diego Becerra de Mendoza and Don Fernando de Grijalva an expedition to an island westward. At sea the two colleagues separated. Grijalva returned to port at Tehuantepec. Becerra was assassinated by his second in command, a Biscayan pilot named Ordono Jimenez. After the crime was committed, Jimenez with two Franciscan monks and some twenty Spaniards disembarked on the shores of a small bay. Here they were killed by Indians to compensate for the honor of putting first foot on the Island of California.

Cortez immediately outfitted another expedition which he commanded in person. After sailing across what was thereafter known as the Sea of Cortez, he landed on May 1, 1536, on this same shore of a barren, deserted land. There was no time to explore it further. News had arrived, calling him back to the capital of Mexico.

What he heard was one of the greatest odysseys told by man.

Eight years before, a vessel of the Narvaes expedition had been wrecked somewhere on the coast of Florida or Texas. Among the survivors were Alvaro Nuñez Cabeza de Vaca, a certain Maldonado, and Dorantes and his black slave, Estevan. Passing from tribe to tribe, suffering thirst, starvation and many privations, they wandered across the continent to the Pacific, turned south and finally reached San Miguel de Culiacan, the outpost of Spanish civilization.

The route of their miraculous journey is not known. But when brought before Cortez, naked save for long matted hair and beards, with feverishly glowing eyes, they told what they had seen. Gleaming cities containing palaces ornamented with sapphires, rubies and turquoise, gold without end. The Seven Cities of Cibola. Cabeza de Vaca swore it was so. And the great black's eyes rolled confirmation.

Cortez had leveled a civilization, gutted an empire of the wealth of Moctezuma so fabulous it had never been measured, won a continent for Spain—and it was not enough. He was in royal disfavor, undermined by rivals, almost stone-broke. To regain his power and prestige, Cortez managed to equip another expedition—by pawning his wife's jewels, it is said.

There were three ships under the command of Francisco de Ulloa. He was instructed to sail north in the Sea of Cortez between the Island of California and the mainland to the approximate latitude of the fabulous Cibola, whence, presumably, it would be simple to march overland. Ulloa set out from Acapulco on July 8, 1539, and eventually found himself at the head of a long gulf instead of in the open sea believed to separate the supposed island from the mainland.

He wrote:

. . . we always found more shallow water, and the sea thick, and very muddy . . . whereupon we rode all night in five fathom water, and we perceived the sea to run with so great a rage into the land that it was a thing much to be marvelled at; and with the like fury it returned back again with the ebb, during which time we found eleven fathom water, and the flood and ebb continued from five to six hours.

The next morning the captain and the pilot went up to the ship's top and saw all the land full of sand in a great round compass and joining itself with the other shore; and it was so low that whereas we were a league from the same we could not discern it, and it seemed there was an inlet of the mouths of certain lakes, whereby the sea went in and out. There were divers opinions amongst us, and some thought that that current entered into these lakes, and also that some great river might be the cause thereof.

These are two notable observations. Ulloa had determined that Lower California was a peninsula, though for a century more it was still considered an island, and Mexican fisherfolk still call the gulf the Sea of Cortez. He was far inland from what is now the mouth of the Colorado, and his observation of a 36-foot tidal rise with its force and fury is the first recording of a phenomenon seen in few other places in the world and nowhere with such ferocity. But as he only surmised there might be a river there, perhaps fearing to wreck his ships by investigating farther, he cannot be considered its discoverer.

Taking possession of the land with formal Spanish ceremony, Ulloa returned down the gulf. One vessel he dispatched back to Cortez with news of his failure to reach Cibola; with the others he rounded the cape to sail up the Pacific coast side and thus vanished forever.

With him vanished Cortez's last hopes. Financially wrecked, rendered politically helpless by the intrigues of his rival, the Conqueror sailed for Spain early in 1540 to plead before his monarch.

The story is told that, unable to secure an audience, he walked up to his sovereign's coach as it passed through the crowded street a year or two later.

"Who are you?" demanded Charles V, looking out upon the unkempt, ruined man.

"I am a man who has given you more provinces than your ancestors left you cities!" answered Cortez proudly.

An unsuccessful way to curry favor with kings! On December 2, 1547, Cortez died near Seville without seeing again the continent he had conquered.

Meanwhile Don Antonio de Mendoza, the viceregal successor to Cortez in New Spain, continued questioning the survivors of the Cabeza de Vaca party. Especially he listened to the great black slave. The Negro's rapt face shone in candelight. Loosened by praise and wine his deep voice shook as he recounted his eight years wandering among the seven golden cities. There seemed no doubt he knew more of the country than his masters.

So Mendoza sent him back to the fabled north he pictured so vividly, guiding Fray Marcos de Niza, a Franciscan monk who had been with Pizarro in Peru. Mendoza's object was simple. The slave would lead the way, and the humble friar would convert the inhabitants to the gentle creed of Christ. Meanwhile the viceroy made plans regarding the gold.

In northern Sonora, Estevan pushed on ahead of the friar and his few attendants. If the reports of the seven cities were indefinite he was to send back a small cross. If what he learned from the Indians seemed encouraging, he agreed to send back a cross "two handfuls long"; while a still larger cross was to be returned if the wealth of the cities was indicated to be great as that of Moctezuma.

Four days later Estevan's Indian messengers staggered back to Marcos's camp carrying a cross as "high as a man."

Marcos hurried forward—always several days behind

Estevan. And now there looms up against the American landscape what is certainly one of the strangest, maddest figures ever seen. The striding black slave known as Estevan, Estevanico or Black Stephen. No one knows whether he was a giant Moor brought from Spain, or an African Negro slave. Indeed, little is known about him at all. He seems a fantastic figure sprung out of the earth to strut across two thousand miles of wilderness and vanish as suddenly and mysteriously.

Only such a stage was big enough to do him justice, and he held it alone. High, dry plains interminably spreading out unbroken as far as the eye could see. Vast deserts burning under a brassy sun. Plateaus slashed with gulch and arroyo, and lifting like altars flat-topped mesas red against the sky. And in the distance black volcanic ridges smooth and stiff as iron, or smoke-blue mountains floating on the horizon.

Against them a great, laughing black slave striding majestically alone.

And behind him, the meek little ecclesiastic trudging in his faded brown robe.

Even today it is a country where swaying sunflowers are mistaken for walking men, where shadows of clouds in the sunshine move slowly over the plains like herds of buffalo, and everything is enlarged and distorted. A country where the illusory effects of light and distance and the clear dry air work perpetual fantasies with the landscape.

Estevan's own fantasy matched that of the country. His was every slave's dream come true. This was his empire and he was its monarch. To the frantic messages of the priest commanding him in the name of the Lord to stop and wait, Estevan paid no heed. Mad with freedom and lust for wealth and power he strode blithely on, a magnificent travesty of his Spanish masters.

For a while he met no opposition. Sending ahead to each

Indian village a gourd decorated with two bells, a red feather and a white, he was received as a healer of the sick and a great medicine man. His black skin, a former sign of bondage, became now the mark of a supernatural man. Soon he was dressed in the finest skins, decorated with feathers, loaded with turquoise and semiprecious stones. Slowly he gathered a retinue of followers from the Indian tribes he passed. And now, entering a village, Estevan imperiously stretched forth a great hand for all that pleased him, food, skins, necklaces of chunk turquoise—women, many women, girls.

Loaded, Estevan moved on. Through the tribes of the Pimas, the Opatas, the Tarahumares and Papagos. Past the little brush wickiups of the Gila. Into the hunting grounds of the Apache and Navajo.

Ahead lay the golden cities of Cibola. According to present Indian legends there actually were six pueblos of the Zuñi Indians: Matsaki and Kiakima, grouped at the base of Corn Mountain; Halona; Kwakina, a few miles downriver; Hawikuh; and Kechipauan, on the mesa to the north.

Estevan grinned—and strode on to his destiny.

No one knows just what happened, where, how or why. But the setting is clear. It seems fairly certain that the city he entered was Hawikuh, the present Indian pueblo of Zuñi. Still it is the largest of the pueblos, its people the most obdurate to change, its ceremonial life the purest and richest. Little wonder that his black majesty, gorgeously bedecked in skins, feathers and chunk turquoise, was met for the first time with a blank look of negation and received a solemn silence in answer to his demands. They are still a proud people, these ancient corn planters rooted to their earth, safe behind thick adobe walls. That terrible Indian negation which has outlasted the centuries! It is worn on their faces yet.

Dumfounded, enraged, perhaps deathly frightened,

Estevan aroused them with his monstrous demands. Instantly he vanished from sight. It is said the Zuñis cut up his black body into small bits which they distributed among the population so as to leave no trace of him. What remains is only a legend among them of a huge "black Mexican" who came their way long, long ago; and the peculiar fact that today, four hundred years later, only Mexicans, of all visitors, are still forbidden entrance to the pueblo.

Stricken with terror, the scrawny desert Indians Estevan had collected along the way fled back to Fray Marcos with news of his death. The simple priest has been called a coward for not eagerly pushing on into Zuñi for martyrdom. He was equally afraid to return to Mendoza without a report of Cibola. So he compromised. He crept up a near-by mountain and stared down upon it.

There it was, flooded by sunlight on the sandy plain. An immense city, large as Seville, with gates of gold, sapphire-studded doors, its walls gleaming in the distance. The fabled city of Cibola as all men wished to see it.

What he saw may well have resembled it. The pueblo was, as its successor still is, the largest in existence, four stories high, and compared favorably in extent with the flat, mud-walled town of Mexico City which then contained scarcely three thousand people. But in that immense flat plain, magnified and distorted by the thin, clear air as much as by his own fear and imagination, the pale-yellow adobe shimmered in the sun; turquoise studded the doorways; then as now bits of straw and flecks of mica in the walls gleamed like jewels. Marcos took one look, mumbled a hasty prayer and fled back to New Spain.

His report threw the capital into turmoil. Mendoza lost no time in organizing a vast expedition. One division was to go by land, retracing Marcos's steps. The other was to proceed by sea along the west coast. They were to keep in communication and to meet farther on.

THE CONQUERORS 139

The expedition by land was headed by Francisco Vasquez de Coronado, governor of New Galicia, the province just north of New Spain. All noble families contributed to its expense; many members, unable to go, equipped and sent servants in order to share in the gold. Pedro de Castañeda, its historian, describes it as the most brilliant company ever assembled. In February, 1540, they started out: three hundred gallants in burnished armor riding horses whose ornate trappings hung to the ground, followed by a horde of servants, men-at-arms and feathered Indian warriors. At the seaport of San Miguel de Culiacan, Coronado turned his back on the last outpost of civilization and struck out north and east.

In May the sea force in command of Hernando de Alarcon arrived at Culiacan. Coronado had just departed, leaving a ship loaded with provisions. Adding this, the *San Gabriel*, to his other two ships, the *San Pedro* and *Santa Catalina*, Alarcon sailed on north keeping a vain lookout for messengers from Coronado.

On August 26, 1540, he arrived at the head of the gulf. Almost immediately all three ships were driven on sand bars "in such sort that one could not help another, neither could the boats succour us because the current was so great. . . . Whereupon we were in such jeopardy that the deck of the Admiral was oftentimes under water; and if a great surge of the sea had not come and driven our ship right up and gave her leave, as it were, to breathe while, we had there been drowned."

Alarcon knew he was experiencing the same phenomenal fury of the waters that had frightened Ulloa. "The pilots and the rest of the company would have had us do as Captain Ulloa did, and have returned again." The fact is that he could not have picked a worse time to encounter the phenomenal tidal bores. It was the last of August, close to the September equinox, when either at the new or the full moon the bores were of maximum intensity. They were dangerous for high-powered craft at any time, but during a month either way from the full-moon bore of September they made passage impossible for any craft.

"But because your Lordship commanded me that I should bring you the secret of that gulf, I resolved that although I had known I should have lost the ships, I would not have ceased for anything to have seen the head thereof."

Alarcon knew also that Coronado was depending on the supplies carried by the *San Gabriel*, and was expecting to meet him any time.

> Now it pleased God upon the return of the flood that the ships came on float, and so we went forward . . . And we passed forward with much ado, turning our sterns now this way, now that way, to seek and find the channel. And it pleased God that after this sort we came to the very bottom of the bay, where we found a very mighty river, which ran with so great fury of a storm, that we could hardly sail against it.

This then marks the discovery by Europeans of the great red river, the immortal river, which Alarcon called El Rio de Buena Guia, the River of Good Guidance, from the motto on the viceroy Mendoza's coat of arms.

Slowly he worked his way upriver by persuading Indians to tow the boats from the banks in the broiling August sun. The Cocopahs he described as naked giants unbelievably strong. One of them carried with ease a log which six of his own men were scarcely able to lift. They reverenced the sun; and on hearing Alarcon proclaim that he had come from the sun and was the son of the sun, accepted for their backbreaking work payment of little crosses made of sticks and paper. After fifteen days of towing, the boats had ascended the river some 85 leagues, and stopped near what is now Yuma at the junction of the Gila.

Here he met Indians who had seen the cities of Cibola

and knew of the black man who had been killed. Not only did the great intertribal highway parallel the Gila River through Arizona, but the desert Indians were wonderful runners and must have carried news of the Negro Estevan through many tribes to the four winds. Heartened, Alarcon ascended the river nearly a hundred miles farther in his small boats.

Still there was no news of Coronado. No messengers arrived, nor did the Quiquimas and Cumanas, possibly Mojaves and Paiutes, allow him to send one overland through their country. At last he returned to his ships and sailed for home.

Coronado and his fellow Spaniards meanwhile had reached the land of Cibola, and from a near hilltop viewed the first of the fabled cities. Writes Castañeda:

When they saw the first village, which was Cibola [Hawikuh], such were the curses that some hurled at Friar Marcos that I pray God may protect him from them. . . . It is a village of about 200 warriors, is about three and four stories high, with the houses small and having only a few rooms. . . .

When they refused to have peace on the terms extended to them, but appeared defiant . . . the Spaniards then attacked the village, which was taken with not a little difficulty, since they held the narrow and crooked entrance.

During the attack they knocked the general down with a large stone, and would have killed him but for Don Garcia Lopez de Cardenas and Hernando de Alvarado who threw themselves about him. . . .

Coronado continued hacking his way into one pueblo after another, looting, raping and spitting their inhabitants on lances. In that year perished the legend of the seven golden cities. He found no jewels, no gold, no ancient cities bulging with centuries of stored-up treasure; only an arid land where water and corn were wealth and its peoples fought hard for survival.

To the south he sent Don Rodrigo Maldonado to search for Alarcon's ships. To the west he sent another officer, Captain Melchior Diaz. And to the northwest he dispatched still another, Don Garcia Lopez de Cardenas.

Maldonado soon returned without trace of Alarcon. Diaz marched on west steadily for 150 leagues, about 412 miles, finally reaching a large red river. On its banks he observed large naked Indians carrying firebrands in order to keep warm, he supposed, though probably to ward off mosquitoes. From this he named it Rio de Tizon, the Firebrand River. Here he learned that it was the river he was seeking, but that Alarcon's ships had just departed. Fifteen leagues upstream he found a tree bearing the inscription: "Alarcon reached this point; there are letters at the foot of this tree."

Digging up these letters, Diaz read that Alarcon had indeed sailed for New Spain. Also that "that sea was a bay, which was formed by the Isle of the Marquis, which is called California, and it was explained how California was not an island, but a part of the mainland forming the other side of the gulf."

Exploring across the river, he came to "sand banks of hot ashes which it was impossible to cross without being drowned as in a sea. The ground . . . trembled like a sheet of paper, so that it seemed as if there were lakes under them. It seemed wonderful and like something infernal, for the ashes to bubble up here in several places."

These were probably the mud volcanoes, geysers and paintpots" of Mullet Island in the Salton Sink.

Turning back, Diaz tried to spear from horseback a dog chasing some of the sheep he had brought along to supply his men with meat. The spear missed the dog, sticking up in the ground, and Diaz was impaled on its butt. For twenty days his men carried him, fighting natives all the time, but he died before they reached Coronado's camp.

The third officer dispatched, Lopez de Cardenas who distinguished himself by his excessive brutality toward the Indians, was under orders to strike out toward a great river of which Coronado had learned. With twelve men he reached a place called Tusayan, some 25 leagues northwest of Cibola. Tusayan has long been supposed to have been one of the Hopi pueblos. It is yet indefinitely located, for the special record of the expedition kept by one Pedro de Sotomayor has apparently never been seen in modern times and is lost in the archives of Mexico or Spain. There is better reason to suppose Tusayan was not one of the Hopi towns; for the Indians said it was twenty days more to the river Cardenas sought, whereas the point he did reach after exactly twenty days is no more than a three days' walk from the Hopi mesas.

The way led through a superb forest and rose up upon the southerly edge of the great Colorado Plateau—"elevated and full of low twisted pines, very cold and lying open toward the north." Then abruptly the earth gave way at his feet. Cardenas was looking into a chasm so stupendous and incomprehensible that he had no words to describe it, nor has any man since found them. The river at the bottom looked like a brook three feet wide, though the Indians declared it to be half a league. For three days the most agile men hunted up and down the rim for a place to descend, but were unable to go down more than a third of the way. Even from here buttes that seemed no higher than a man were found to be taller than the great tower of Seville. From rim to rim at the top of the gorge, the width of the cañon appeared to be three or four leagues, eleven miles or more by air.

Cardenas had discovered the Grand Cañon of the Colorado.

Castañeda, chronicler of the Coronado expedition, reported that the river therein was the same Tizon but "much nearer its source than where Melchior Diaz crossed it."

Cardenas gave up trying to explore such a mighty chasm and returned to the main body of the expedition. Coronado now struck out again behind a new Indian guide, an accomplished liar nicknamed the Turk, toward the fabled Gran Quivira whose emperor was sleeping under the tree hung with golden bells. They traveled east and north in a great circle that probably included the vast grassy plains of Kansas and Colorado, perhaps touching Missouri. There was nothing but a prairie flat and limitless as a sea, darkened by herds of buffalo. The soldiers, starved and mutinous, finally rebelled and strangled the Turk in his sleep. And then in 1542 began the long straggling march back to Mexico City devoid of gold and jewels, without honor—without illusions.

After Coronado no more is heard of the wonderful Isle of California, of El Dorado and the Gran Quivira, of the golden Seven Cities of Cibola. California had been proved a peninsula and not an island. The mighty red river had been discovered and ascended nearly 234 miles from its mouth. Its strange tidal bores had been experienced and recorded. Its upper reaches had been visited, the Grand Cañon seen for the first time by a white man. All this within a scant six years, less than thirty years from the time Cortez stepped off his ship on Aztec earth. Yet for half a century thereafter the Colorado River lay disregarded.

Coronado had served the real purpose of all explorers, to pierce with his lance the shell of illusion to the kernel of truth within.

But that truth, stranger than all illusions, was too great to believe. The conquerors, paradoxically enough, were conquered by their own lack of imagination.

CHAPTER 2

The Padres

ALL THE conquerors brought along a priest or two on their marches. With the sword of destruction came the cross of salvation. It was not enough to massacre the native populations; their souls must be sent to God. And so on every mountaintop there loomed the stark symbol of crucified man, arms outstretched against the sky.

Today in most of the pueblos a cross still stands in the plaza. On saints' days an early Mass is said in the little adobe church, behind. But quickly as it is over and the priest goes home, the people strip and paint, filing out into the plaza to the low beat of a drum. There they dance as they have always danced. For gently falling rain and growing corn. To the ancient gods of earth and sky, the eternal gods which endure through all changing creeds and religions.

The creed of the cross did not change the people, but it did change the face of the land. Ever after it marked the steps of a breed of men who followed in the wake of the conquerors, and who for nearly two centuries were the only ones to cross the vast basin of the great red river of the West.

Preliminary to these *entradas* there were several attempts to settle the country and Christianize the people. As early as 1583 Don Antonio de Espejo traveled to Zuñi and the Hopi towns, followed by concerted attempts to subdue

Acoma on its lofty clifftop. Like the searches for great wealth they were failures. There was always something inherently inimical in the very landscape to the Spaniards. For long they dominated its peoples, yet at the last it was the land that drove them out.

But from 1604 until 1781 it belonged to the padres. It was a remarkable period and they were remarkable men—these sturdy, brown-robed priests who recognized in principle only the domain of the spirit and were engaged in mapping out a wilderness empire in fact. For to the east, along the Rio Grande, were being established the first northern outposts of New Spain. And to the west, along the Pacific coast, was being founded another long chain of missions and presidios. It was the necessity for finding a route between these two which impelled the prime entradas of the padres.

During this period two major catastrophes befell the padres. In 1680 occurred the famous Hopi and Pueblo uprising, in which all the missionaries were killed. And in 1767, on an order of the Pope occasioned by a change of European politics, the Jesuit Order was suppressed. The six thousand Jesuit priests in Mexico were replaced by Franciscans. The change was sudden and unexpected. The Jesuits had laboriously founded and constructed a chain of fourteen missions up the whole barren length of Lower California—missions whose ruins still today await rediscovery by historians. Now, with but twenty-four hours' notice, they were ordered to abandon their hard-won flocks and handiwork, and retaining only their crosses and Bibles, to flee the country.

The Franciscans who replaced them added only one mission to the chain. They found the long desert peninsula too frightful to endure. The missions began to crumble; the Indians fled to the hills. Strangely enough, the Lord at this time directed the padres to new labors northward, leaving the peninsula to the Dominicans who built seven new missions.

The new field for the Franciscans happened to be a beautiful coastal plain of gently rolling hills perpetually green, mild of weather, and inhabited by timid, backward Indians. So here, from San Diego to San Francisco, grew up a new chain of twenty-one missions whose very names evoke a romantic softness, the chime of bells at twilight, the fragrance of orange blossoms, long shady portales and lovely gardens—Mission San Juan Capistrano, San Gabriel Archangel, Purisima Conception and San Luis Rey de Francia.

The Colorado River basin was something else, far sterner. Yet despite these setbacks, the work continued even here. Of all its padres whom history has recorded, scarcely a dozen stand out; only three were truly famous. A handful of unarmed men pitted against an unknown wilderness of snowy peaks and blazing deserts larger than all Spain and Portugal together. Little wonder that their steps have trodden deep into our maps and memories.

In 1604, Don Juan de Onate, the first governor of New Mexico, began the first entrada with a notable company of two padres and thirty soldiers. They left San Gabriel de los Españoles on the Rio Grande—the first town founded in New Mexico and the second in the United States—in October, and struck out westward.

The seven Hopi towns on their high isolated cliffs loomed first out of the pelagic plain. For Onate, as for Espejo before him and for every traveler after him, they were the one oasis and most important landmark in the region. The people were known as the "Moquis." By that name they are identified in all the old records, and hundreds of pages since have argued the later impudent use of "Hopi" instead. But Hopi is the word; it is a contraction of the people's own name for themselves, "Hopituh"—People of Peace.

Ten leagues beyond, the company crossed a river which from its muddy red color Onate gave the name of Colorado. This was the first use of the name, but the river was only the

Little Colorado. Believing that it discharged into the "South Sea" (the Pacific), the company descended the Colorado Plateau to the headwaters of the Verde. Here appeared Indians whom the soldiers called "Cruzados" from the little crosses they wore, perhaps gifts from Alarcon years before. They directed Onate to another small river flowing into a large one which in turn flowed into the salt sea twenty days distant.

The small river, probably Bill Williams Fork, Onate named San Andreas. The large one was the Colorado at last. In high spirits, Onate called it the Rio Grande de Buena Esperanza, the River of Good Hope. Down it they marched to the Gila, giving this the name of Rio del Nombre de Jesus. And on January 23, 1605, they reached the mouth of the Colorado.

The way back led over an already ancient trail from the river to Zuñi, the fabulous Cibola, south of the Hopi pueblos. Near by, between the cedar cañons, they came to an immense sandstone cliff rising over two hundred feet high, like a lofty tower with white battlements. From its striking resemblance to a Spanish castle the soldiers promptly named it El Morro. On its sheer wall, in old Castilian, Onate proudly inscribed the following legend:

> Passed by here the Commander
> Don Juan de Onate from the
> discovery of the South Sea
> on the 16th of April, year 1605.

This carving set a precedent for the travelers who passed there during the next four hundred years. Every name of note was inscribed there as well as the modest two lines of a poor, common soldier, "I am from the hands of Felipe de Arellano, on the 16th of September, soldier."

Nine days after making the first entry on this magnificent Inscription Rock, Onate reached Española. He had completed the first entrada of three thousand miles on foot in not quite seven months.

Sixteen years later Zaldivar with Fray Jimenez and forty-seven soldiers attempted the same route. Fifteen leagues west of the Hopi pueblos, confronted by gigantic cañons, they turned back. And now for another lapse this wild cañon region of the Colorado was abandoned. The Hopis and Pueblos suddenly rebelled. In one day twenty-one missionaries and four hundred other Spaniards were killed, and the others driven back into Old Mexico. According to the Hopis, their death atoned for the gift of Hopi girls given by the priests to Spanish soldiers. For nearly a century and a half no stranger's foot halted on the brink of an unknown chasm, no white face peered through the cedars.

When next one appeared, it was only in the deserts to the south, and he was the first of the three most famous padres. Eusebio Kino (Eusebius Kuehne) was an Austrian Jesuit. From the little mission of Dolores in Sonora, which he established in 1687, he traveled over the whole of Sonora and the half of Nuevo Mexico now known as Arizona. Pimeria Alta, the upper land of the Pimas; Papagueria, the land of the Papagos; the deserts of the Coco-Maricopas, Yumas, Mojaves, Cocopahs; and the wild sierras of the Apaches and Yaquis—it was all one region of vast yellow silence. Deserts of drifting sand, jumbled hills of sun-blistered rock, devils' gardens of giant cactus: a region where water holes were days apart and landscapes swam upside down in the quavering heat waves.

But Kino saw straight, through every mirage. His map of 1701 is a positive delight. Sonora then was the starting point for the unknown, vacuous north. Today this north is but yet an infrequent starting point for the bleak Sonoran mesa and the remote Sonoran hills. So that on Kino's map one will find the names of small, forgotten Indian settlements scarcely known today.

There are shown six of the seven legendary Yaqui villages: Rahum, Potam, Vicam, Torin, Bacum, Cacorim. There—look!—is Tonichi (he calls it Tonici) up the Yaqui River. But first you must get off from the main train that runs three times weekly across the lower river bottoms and sleep overnight in a thatched ramada beside the solitary boxcar station. Next day, all day, a weekly two-car spur train with ten seats for passengers crawls up the tortuous Yaqui River gorges. At twilight it stops dead on a crude round-table on top the bluff. So you get off in the strange effulgence, in the thick chaparral that swallows up the horses waiting for the other two passengers. Trudge down the trail to the river and wait hours for an Indian boy. Kneeling on his raft, you watch him swing you out midstream, then neatly on a twist of the current pole safely into shore. Climb up a winding trail to the top of the opposite bluff. To a dark, deserted plaza surrounded by squat adobes. To a great stone-blackened church staring off into the darkening, empty hills. Tonichi! How the devil did one get there even twenty years ago, to sleep on the earthen floor of a hut, after a swig of bacanora?

Bacanora: there it is too, a clump of adobes whose residents conjure from cactus juice that clear white firewater famous among neighbors who wear handmade leather vests and chaps against the cholla, and ride a hundred kilometros for a copita to warm their insides. Kino spells it "Babiacora."

Chinipas (without the s) in the old mining district is shown, where men still pull out old Spanish ladders with rawhide fastenings from the caved-in workings.

There is Cumuripa too, spelled "Comoripa." The "Widows' Town" I heard it called when I rode through just after the 1924 Yaqui uprising when so many of its men were killed.

And Sahuaripa. Where is that? To hell-and-gone since Padre Kino vanished. One rides saddleless days on end, living on dry tortillas and the savage little wild pigs that come bursting out of the chaparral, and the surging gray hills keep heaving up into the Sierra Madres bestriding Sonora and Chihuahua.

This was Padre Kino's life for almost thirty years.

On top of one of these hills he saw the true outline of the gulf at the mouth of the Colorado, and thereupon drew his map—the first to show the head of the gulf correctly. He was the first white man in a hundred years to reach the mouth of the river, and the first to see the ruin of Casa Grande, often called the House of Montezuma. The Gila he named the Rio de los Apostoles; and the Colorado, Rio de los Martires. Before he died, Kino selected the site of San Xavier del Bac, south of Tucson, and began the church which remains one of the most beautiful Spanish-Colonial missions, and the only one in the Colorado River basin.

After Padre Kino's death in 1711, there were few hardy enough to venture into his domain. Padre Jacobo Sedelmair in 1744 went down the Gila to the great bend and cut across to the Colorado. Two years later Fernando Consag sailed up the gulf to the mouth of the river looking for mission sites, and sailed hastily away. Don Juan Maria de Rivera in 1761 went north as far as the junction of the Grand and Gunnison, crossing the divide at a place he descriptively called "Purgatory." Fray Alonso de Posadas reached the other branch, the Green, which he named the San Buenaventura. These entradas opened up again the mountain hinterland abandoned by Zaldivar 158 years before.

This old trite phrasing is peculiarly apt. After every entrada the earth, fold on fold, range after range, closed up again. Already the Colorado had been given four names —Rio de Buena Guia, Rio del Tizon, Rio Grande de Buena Esperanza, and Rio de los Martires. No man seemed to remember those who had gone before, or the routes they so blindly followed. It was as if no one yet had found the key; and so they sought, one after another, to open up an earth that was forever closing after them, remaining ever new and still inscrutable.

Now, at the end of the period, entered the two padres who trod deep and left their trails to be followed ever after. Both were Franciscans in name and order—Francisco Garces and Francisco Silvestre Velez de Escalante.

Father Garces left a biographer: his chaplain, Padre Font, who wrote a diary. In it we see him as a zealous missionary complaining of a sickness of heart at the contemplation of the souls that must inevitably be lost because he had failed to sprinkle the necessary three drops of water on their dark skins. Dr. Elliot Coues wrote of him that he lived the life of Christ. Certainly he was a robust crusader who endured every hardship to reach even the most remote and inaccessible converts.

Garces made his headquarters Kino's mission of San Xavier, finally finished in 1797. A new presidio at Tubac, near by, gave it military protection. From here, beginning in 1768, Garces made five entradas. He was the first to make regular use of the name "Colorado": applying it, he explains, because the river, draining a red country, was tinged red in the month of April when the water was high from melting snows. The padre was a careful observer; he showed the first perception of the river's enormous drainage area.

> This much is certain, that from the Yutas [the Colorado Utes], who are on the north of the Moqui, until its disemboguement into the Golfo de California, it gathers to itself no notable body of water; wherefore it is likely that the greater part of its abundance comes from far beyond.

He also advises that the Colorado was called "Javil" by the Yumas, and was a translation of the Piman "Buqui Aquimuti."

The fourth entrada was a remarkable journey. With Captain Juan Bautista de Anza, Garces crossed the Colorado, and then struck across the Colorado Desert to open a route for communication with Fray Junipero's new mission at Monterey, California. At the mission of San Gabriel, near Los Angeles, he left the company and returned alone. The route is easily followed. Every step of his journeys, even his famous fifth entrada, can be traced from his accurate, topographical descriptions. The river crossing was made at Yuma, and the desert crossing through the upper end of Imperial Valley past the Chocolate Mountains and the Chuckawallas. It was a spectacular, frightening journey, yet Garces immediately wrote his contemporary, Escalante, in Santa Fe, that he had found the best route to the Franciscan missions in California.

Next year, in 1775, he joined the memorable expedition made by de Anza to found a new mission and settlement on San Francisco Bay. What a caravan, to ford the Colorado and plod 1,500 miles without the loss of a man or beast: 240 soldiers, muleteers, Indian interpreters, women and children; 165 pack mules loaded with ammunition, baggage and presents for Indians; 530 horses, mares and colts; 355 cattle!

But at the Colorado, Garces left it. Thus begins his own fifth and famous entrada.

From the Mojave Desert to the San Bernardino Mountains, he circled back by way of Lake Tulare and Kern River. Thence alone except for some Wallapais guides, he struck eastward from the Mojave to the little 400-acre retreat of the Havasupais deep in the cañon of Cataract Creek. The trail down was so steep it was impassable for his mule; he had to use a ladder for part of the 2,000-foot descent.

He was the first white man to see the Grand Cañon from the west, possibly from the site of El Tovar. The Little Colorado he called Rio Jaquesila de San Pedro. Finally, on

July 4, 1776, he sighted the familiar landmark of all travelers: the Hopi mesas.

Across the continent on the opposite coast from the one on which the de Anza expedition was settling, a new race of intruders was celebrating its first Independence Day when Garces rode into Oraibi. The weary padre descended from his mule, expecting food and solace, offering salvation and gifts. The Hopis offered nothing, accepted nothing. They simply bade him leave. They too, in this high hinterland of America, were insisting on their own traditional independence.

Padre Garces jogged wearily back to San Xavier, arriving September 27, 1776, and completing five entradas that covered at least five thousand miles, through nine tribes and twenty-five thousand Indians.

Four years later, across the river from Yuma at the site where the de Anza expedition crossed, a combination mission and presidio was established: Puerto de la Purisima Concepcion. Another eight miles down was also founded San Pedro y San Pablo de Bicuner. Padres Garces and Barraneche were in charge of the Immaculate Conception mission, and Padres Diaz and Moreno at the St. Peter and St. Paul. Each was protected by a few soldiers and a nucleus of colonists who had taken from the Indians all the best surrounding land.

For a year the Yumas waited, stifling their resentment. The next summer Captain Moncada, lieutenant governor of Lower California, arrived with a company of soldiers and recruits bound for the California mission settlements. They pastured their horses on the fields of mesquite beans on which the Indians depended for food. That night, on July 17, 1781, the Yumas struck. They were led by "Captain Palma," a Yuma Indian who had been Padre Garces' convert and protégé. All the Spaniards were killed, including Garces; and this ended all the missions along the Colorado and the period of the padres.

Meanwhile back in Santa Fe, Garces' contemporary was chewing the edge of his robe. It was 1776 and a worried

group had gathered to discuss the best route to the mission at Monterey, California. There was Don Juan Pedro Cisneros, mayor of the town of Zuñi; Don Bernardo Miera y Pacheco of Santa Fe; Fray Francisco Atanasio Dominquez, a visiting delegate, and five others. Their leader, and hence the most concerned, was, as he describes himself in his diary, Fray Francisco Silvestre Velez de Escalante, teacher of Christian Doctrine in the Mission of Our Lady of Guadalupe of Zuñi.

Garces' letter to him had not yet arrived saying that he and de Anza had found the best route to Monterey.

Padre Escalante after some deliberation rejected the southern route, believing a northern one would prove better.

Accordingly, on July 29, 1776, the party of nine left Santa Fe and headed north: the two padres, the alcalde of Zuñi, the military captain of Santa Fe, and five soldiers. They marched steadily northwest up the Rio Chama and crossed the San Juan on the New Mexico-Colorado line. Cutting across the southwest corner of Colorado, past Durango, they went down the Dolores for eleven days and headed due north through Colorado. Crossing the Colorado just west of Grand Junction, and then the White, the party turned west into Utah and crossed the Green south of the Wyoming line. Through the Uintah Mountains, over the Wahsatch Range, they came down the western slope by way of the Spanish Fork to Utah Lake near Provo.

It was now October 8th, and in his diary Escalante wrote:

> The winter had now set in with great rigor, and all the mountain ranges that we could see were covered with snow; the weather was now very changeable and long before we could reach them [the settlements in California] the mountain passes would be closed up, and we would be obliged to remain two or three months on some mountain, where there were no people and where we would not be able to provide necessary food. The provisions . . . were now nearly exhausted, and if we continued to go on we would be liable to perish with hunger if not with cold.

Still it snowed, and the lofty peaks gleamed whiter. The men continued to grumble. That night Escalante proposed that they all cast lots to decide the route to be followed. To this:

> . . . they all agreed like Christians, and with fervent devotion recited the third part of the rosary, while we recited the Penitential Psalms with the litanies and the other prayers which follow. Con-

cluding our prayers, we cast lots and it came out in favor of Cosnina. We all accepted this, thanks be to God, willingly and joyfully."

It was a significant decision. Cosnina was the name which Escalante used for the Colorado—Rio del Cosnina; and its selection meant that they had decided to give up their attempt to continue west to Monterey, and to turn back south.

Wherefore Escalante's entry of the 9th is headed "New Route, and the Beginning of Our Return." He felt a certain disgust "on account of our abandoning the route to Monterey to follow this one." This he alleviated with the belief that God had had a hand in the casting of their lots for their return "that we now understood to be expedient, and according to the Holy Will of God, for Whom only we desired the journey, for Whom we were willing to suffer, and if necessary, even to die."

The Holy Will cannot be considered as having erred in judgment, but it did let the lot casters in for geographically needless trouble of the worst sort. For if the party, at the point where it first turned south, had continued due west according to the original plan, it would have struck the comparatively easy trail straight across Nevada to Carson City, around Lake Tahoe to Sacramento and thence to San Francisco—the same route followed years later by the Pony Express. Instead, the party plunged headlong into the labyrinth of cañons north of the Grand Cañon.

The route is too long and the hardships too many to be followed except in Escalante's own words. They struggled due south through eastern Utah, crossed the Virgin River into northwestern Arizona, and then turned east across the Kanab. Food ran out in the first few days, and from that time on they subsisted on pine nuts, roots, herbs, and "a few toasted cactus leaves." When these gave out, they began to slaughter their pack horses.

Early in November they reached the Colorado and finally found a crossing. It was just north of Marble Cañon, above Lee's Ferry, and thereafter became known as the "Crossing of the Fathers." On the morning of November 7th, they began to ford it. Steps had to be cut into the cliff to get the horses down. The baggage was lowered over the bluff "with ropes and reatas, down to the vicinity of the ford." At "above five in the evening we finished crossing the river, singing praises to God, our Lord, and discharging some muskets in sign of the great joy that we all felt at having conquered so great a difficulty." They were the first white men to cross the Grand Cañon, a feat roughly contemporary with Washington's tamer crossing of the Delaware. Of the ford which they appropriately named "La Purisima Concepcion de la Virgen Santisima," Escalante adds this: "Here the mountain sheep [bighorns] breed in such abundance that their tracks seem to be of great flocks of tame sheep. They are larger than the domestic sheep, of similar build, but much swifter."

Thence the route back to Santa Fe lay southeast to the Hopi mesas and over the familiar trail followed by Tovar, Espejo and Onate. Escalante arrived on January 2, 1777, and sat down at once to write his diaries from his notes, which were delivered to the viceroy in Mexico City within a few months.

Thus the period of the padres. Of it nothing was left except the deeply engraved routes of their remarkable entradas. From the cathedral in Santa Fe, New Mexico, to the missions in California, from San Xavier in southern Arizona to Wyoming, there has never been a true large mission in the Colorado River basin. The only possible exception was the beautiful church at Acoma on its lofty clifftop; and it was not a mission, having no settlement and protective presidio, nor a docile congregation supporting it or its faith. Carried up the 357-foot cliff, stone by stone, beam by beam, a handful of earth at a time, it was a labor paid for by the most consistent and bloody rebellions that occurred in any of the pueblos. San Xavier alone, near the Mexican border, remains the Colorado's only monument to the padres.

Two hundred years of almost superhuman effort by truly remarkable men; backed by the political and military power of a nation dominant in both hemispheres; directed by the greatest religious organization in the world; and presumably sanctified through it by God Himself—and yet completely futile. Fruitless today, almost another two hundred years later.

It is not too early or too impertinent to inquire briefly why.

There are two answers, two viewpoints. One we know: sour grapes. The half-naked savages were too undeveloped and uncivilized to understand the saving word of grace.

Now really what did even the learned Jesuits have to teach? In essence it was an ecclesiastical doctrine already so intellectualized and dogmatized that it could not answer the simplest irritating questions. If the white man's God was everywhere, why was He not in the corn plant and the mountain peak? (Not to ask why it required an edict of the pope to grant Indian bodies official souls.) If His realm was wholly of the spirit, why did the Franciscan, Dominican and Jesuit fathers themselves quarrel over temporal boundaries? The Devil himself prompted the questions of where man came from and what was his relationship to the universe around him.

Do not doubt that the Pueblos and Navajos asked. And to them the single angry shout or despairing cry of "Faith!" —the only answer—was not enough. Silently, with the patient tolerance of a long-enlightened people for ignorant barbarian guests, they politely accepted the three drops of water sprinkled on their heads and went back to their own belief.

Even while the padres were wearily trudging from question to question, their superiors were deciding that the creation of the world had taken place in the year 4004 B.C., on the twenty-sixth of October, at nine o'clock in the morning. This time, set by Archbishop Ussher, was inserted in the authorized version of the Bible, and received the sanction of the Christian Church. Neither he nor the padres knew that there existed a faith here whose historical framework alone reached back millions of years before this earth itself was created.

There have been four successive worlds. There was the First World, the Dark World, the Running-Pitch World. There was the Second World, the Blue World, not yet lit by sun or moon, and bathed only in the blue luminous light of the shining mountains. There was the Third World, the Yellow World. And there was the Fourth World, the White World, the Present World, covered long after its creation by a flood lasting forty-two days.

From one world to another the people rose in successive emergences, led by the six sacred personages including First Man and First Woman, prototypes of present man. Chief of all was the great Creator, whose name literally translated means "The-Love-a-Mother-Gives-Her-Child." The transplanted mountains still remain chronologically named: the Sangre de Cristos, the "Third Mountain in the Third World"; Jemez Range, "Fourth Mountain in the Third World"; and the Chuskai, Black and other ranges westward toward California, mountains made in the Fourth World. The Place of Emergence was in the southern Colorado mountains near Silverton, geologically the oldest dry land on this continent. Always this ancient homeland of the Navajos has been bounded by the San Francisco peak of Arizona on the west, the Colorado mountains to the north, and the sacred peaks of New Mexico to the east and south.

Thus relates the Navajo Creation Myth, and to the padres the gibberish heathen chants they heard nine nights in a row under the frosty stars were the 800-odd songs that told it in full, memorized detail—told of the creation of man and mountain, of all life, of the period of prehistoric monsters whose footprints are still embedded in rock, of the petrifying forest—told it in great singing poetry, the only indigenous poetry of America.

Deep in their underground kivas, the Hopis, Zuñis and Pueblos orally handed down their own ritual poetry from generation to generation. Literally the voice of the rocks themselves. Their sacred traditions varied little from those of the Navajo. There were the same evolutionary four worlds symbolized by color, the six sacred directions—east, south, west, north, zenith and nadir. And each of these four worlds signified the four primary elements through which all life progresses, and of which man is composed—the first world of fire, the second of air, the third of water, and the fourth of earth. As in the Navajo, did not the last people made have the name of "Made From Everything?" Do not our own Nebular and Planetesimal Hypotheses conceive of planetary creation as cooling from a fiery mass, the escaping oxygen forming vaporous air, condensing into water which finally separated from the earth in great seas?

The Pueblo kiva, seen from inside, like the Navajo hogan, was the world itself in miniature. Its round walls showed the circular horizon. The small sipapu in the center of the floor, and the single ladder—opening above, were themselves the places of emergence. Lit by the parent fire, the symbolic stylized designs of corn meal and pollen shone out rich in color, composition and meaning to illustrate the chants.

Believing that their own evolution was taking place constantly and coincidentally with that of the universe about them, they were a people deeply conscious of the presence of life in the corn plant and the mountain peak alike. One could not develop without the other. Nor could they themselves without both. Hence within this framework of universal evolution lay the core of their faith—the achievement of a harmonic relationship between man and his universe. This relationship was what the Navajos called "hozhonji," or happiness, and whose effort to achieve it the Zuñis called "The Road." It was all based upon the fundamental that even a mountain has a spirit form as well as a material one; that all matter has a spiritual essence which cannot be separated from its material composition.

Hence the use of the sand paintings for curative purposes. Colored sand was obtained from the mountains corresponding to the chant, made into great flat paintings of sacred stylized designs, then sprinkled over the patient's body and immediately destroyed. It was not wholly that a man's body was ill from a snakebite or an enemy wound. By the contributary cause of his own action he had introduced disharmony in his right relationship to the universe, and must be psychologically put back on the road as well as therapeutically cured.

It is strange that in all the world there exists probably but three places where such deep-rooted convictions of such wide and advanced scope have persisted for centuries: in the Himalaya Mountains of Tibet, in the Peruvian Andes, and here in the high plateaus of the Rockies. The similarities are many. Not only do the Tibetan mandalas compare to Navajo sand paintings, but the Buddhist "path" compares to the Zuñi "road." They do not imply a false ethnological derivation either way. Rather an ultimate source that must exist hidden behind all lesser conflicting creeds. Both coincide in a system that, within an evolutionary scheme great enough to include all forms of life, recognizes the essential embodiment of all materiality with a corresponding spirituality.

Against this the padres were powerless; at best, they could perceive in it nothing but a limited animalistic belief, if that.

But it was a significant first meeting of two races which were not only oceans but civilizations and psychical poles apart. One of them was a bluff, active, outward-thinking race, accustomed to force and dominance, and whose symbol was the phallic spire of conquering man. It believed in a vast materiality that was to build up here the most stupendous mechanistic-mental civilization ever seen on the face of the earth.

The other believed in substance above form. It was a race silent, secretive, introspective, with its symbol the circular womblike kiva sunk deep in the fertile, unresistant earth. A people who still believe in the strength of the spirit that impenetrates even stone, the spirit that in time moves all mountains.

Not for nothing did the last of the padres turn silently away. Through the long blue shadows purpling on the plain, and into the memory of man, the people watched him go. Those old, dark Indian faces, wrinkled like leather, which having seen four worlds and all their changes could still envision perhaps a fifth, of which they too would still be an inalienable part.

CHAPTER 3

The Trappers

A FEW YEARS after the last of the padres, a stylish young blood in far-off London smirked into his mirror as he put on his hat. He had been born with a silver spoon in his mouth about the time that Father Garces achieved martyrdom. He had attended Eton and had linked arms with the young Prince of Wales, later King George IV, who had given him a captain's commission in his own royal regiment. Now at the age of twenty-one he already had left the service, inherited a tidy little fortune of £30,000, and set himself up in a new bachelor establishment in fashionable Mayfair. Egad! What a dashing chap he saw in the mirror, dressed up like a Thanksgiving turkey: Beau Brummel, already recognized as *arbiter elegantiarum* throughout the world of fashion. With his right hand he gave a last deft flick to the brim of his new hat—and with that one gesture this addle-headed English fop accomplished more than all the padres had in the last two hundred years, and the Spanish conquerors in the century before them.

For wherever he now strutted, into the smoking rooms of Mayfair, through the streets of London, along the boulevards of Paris, it was to the ecstatic refrain of "Where did you get that hat?" It was indeed a beautiful stovepipe, nobly built as a queen, and made of the finest beaver. Every man had to have one just like it.

Behind that hat lay a long tradition. In his own elaborate book, *Male and Female Costume*, Beau Brummel wrote that hats were first worn in the time of Henry VII, who was crowned in 1509. But the Beau himself did not know just when the skins of beaver had been first used in hats, save that they were first imported from Flanders during the fourteenth century. In Chaucer's *Canterbury Tales* the merchant is described as wearing

> on his head a Flaundrish beaver hat.

By the seventeenth century the beaver hat was in vogue. The reigns of Louis XIII of France and Charles I and Charles II of England are associated with the great widebrimmed beaver hats so beautifully painted by the Dutch master, Frans Hals, and his Flemish contemporary, Sir Anthony Van Dyck. A song of the time bears testimony to the value of the English beaver:

> The Turk in linen wraps his head,
> The Persian his in lawn too,
> The Russe with sables furs his cap,
> And change will not be drawn to;
> The Spaniard's constant to his block,
> The French inconstant ever;
> But of all felts that may be felt,
> Give me your English beaver.

Pepys in his diary of June 26, 1661, writes:

> This day Mr. Halder sent me a beaver which cost me £4 5os—an enormous price for a hat considering the value of money at this period.

European beaver was almost exterminated to supply the demand. English beaver became no longer beaver, but a substitute mixture of lamb's wool and rabbit's fur. Then gradually as real beaver came back, the quality improved and prices rose. Writes Philip Stubbes, able chronicler of his day: "Some are of a certain kind of fine hair; these they call *beaver* hats, of twenty, thirty and forty shillings apiece, fetched from beyond the sea, whence a great sort of varieties do come."

What had happened was the replacement of the former European beaver with American beaver. By the end of colonial times beavers were almost exterminated throughout the eastern colonies. The hunt continued in Canada, largely through the Hudson's Bay Company and its rivals, and extended finally into the country west of the Mississippi.

The stage was set. By the beginning of the nineteenth century the wide-brimmed hat was ready to give way to the tall tylindrical form of the "stovepipe," championed by the Beau. And in America was a new supply of beaver. Excellent beaver. In fact, the first tall "silk" hat later made in Florence, Italy, was a beaver so highly polished to a satinlike luster that it was called a "silk beaver."

Wherefore in 1799 Beau Brummel ushered in a new style and a new century, and founded in the vast basin of the Colorado the Beaver Hat Empire.

For here in America were to be found the millions of beavers for millions of hats. Everywhere—wherever there were water and trees with an edible bark. From the western edge of the Great Plains to the Pacific coast; from the Arctic Circle to the Mexican border; from the high peaks at timberline down to the muddy creeks sinking into the deserts. But in the hinterland along the Colorado they were thickest, and they still are. There is scarcely a boy in this region who does not boast of having spied upon them at work on their dams, heard the warning plop of their flat tails when he made the slightest move.

So here rushed the trappers to trap them. What buffalo were on the plains, beaver were in the mountains. Eighty skins to a pack, weighing one hundred pounds, and worth five hundred dollars—and sometimes fetching as high as fifteen dollars apiece—beaver became the regional currency. Against them powder and lead and salt were weighed and measured. Built on beavers, the great fur kingdom was established overnight and spread to the most remote borders.

In just forty years its reign was over. And when it was over, the country was no longer what it had been. There was not a stream, a valley, a mountain ridge that was not known. Time itself had been changed. It was to be counted no longer in centuries, but in decades. And none of this really mattered. The really significant thing was the creation of the breed of men who achieved all this. They belonged to no race, no creed. Unique in time and space, deeply prophetic, they were born of the immensities of the American wilderness, in the travail of its stupendous solitude. They belonged wholly to the mountains. Trappers, traders, scouts, hunters, guides—whatever they were and later became, they were all mountain men . . .

For three centuries all the strangers who had penetrated the Colorado River basin had come from the south. But now, by 1800, the dominance of Spain in the Old World had been broken, and the Spanish empire in the New World, like a vast sea that had washed northward, was now slowly receding. A new nation, lusty in its youth, had grown up on the Atlantic seaboard. So it was that from the east these new strangers came, just as the first had come to the New World three centuries before.

The Missouri was the western limit of this new civilization. The men who left it behind them faced an uncharted grassy plain almost a thousand miles wide and extending the length of the continent. Over it raced the fast-riding, nomadic tribes, Osages and Sioux, the Crows, Cheyennes, Arapahoes, Comanches and Kiowas, thundering down swift and sudden upon the slow wagon caravans to protect their tribal hunting grounds from invasion. By the time one of these strangers had crossed it, he was no longer the man who had left the far Missouri. His small wagon train had disap-

peared, most of his comrades, all of his supplies. At best he had left him a mount, a string of steel traps, rifle, powder, lead and salt. More often he found himself with nothing but his knife. But he considered himself lucky to have only his scalp.

And now he saw rising before him something he had never seen before, something that would haunt him always, something of which forever after he would be an inalienable part. The mountains! That great front wall of the Rockies looming into the clouds above him, extending to each side

THE TRAPPERS

as far as he could see, from Canada to Chihuahua. Alone, he crept warily into its blue shadow, was swallowed by its black cañons.

A year, two years, later he emerged—if he did emerge —a man completely different, entirely new. He was dressed in a greasy, bloodstained, buckskin shirt and leggings, wearing Indian moccasins. His body had hardened to unbelievable endurance. He had learned to live on meat—raw, rattlesnake or even Indian flesh. Salt was a luxury. A precious pinch of tobacco lasted weeks when mixed with the inside bark of a red willow and smoked sparingly at night. A buffalo robe or a bearskin was enough of a covering even in snow. When he slept, it was warily like a beast, apart from his tiny fire. At the slightest sound his body sprung—like one of his own traps—hand to knife.

To be sure, he had his bales of beaver. What were they? The hundreds or the thousands of dollars they were worth meant nothing, nor did the comforts they might have bought. He got rid of them as fast as possible. No longer could he endure companionship; he had forgotten the sight and sound and proximity of mankind; he had almost forgotten how to talk.

Only one thing possessed him. To flee all that which had brought him back across the plains. To get back to danger, privation and solitude, to get back into the mountains. And so one night he suddenly vanished. This time he went alone.

Soon there was no need for him to return at all. There appeared in the mountains themselves trading posts where he could rid himself of his skins in exchange for new supplies.

Behind them were growing powerful fur companies hiring hundreds of trappers and dominating great areas: the Hudson's Bay Company; the American Fur Company, Astor's preliminary venture in 1808, followed by his Pacific

Fur Company; the Missouri Fur Company, established in 1810; the Northwest Company; the Rocky Mountain Fur Company. Their owners became veritable kings. Out of beaver John Jacob Astor built one of the greatest fortunes in America. Lisa became a legend, Ashley famous. Like the fabulous Choteau, some of them took Indian wives in order to make friends with the tribes whose furs they coveted. Most of this fur trade finally centered in St. Louis, which thenceforth became the fur center of the continent.

But the trappers were too aloof and independent to stay long in company employ. Most of them remained "free" trappers. So, despite these company posts, there grew up a few great trading centers which became the principal rendezvous for all the mountain men—traders, company and free trappers alike. They were like nothing that ever existed; not to be compared with the rude military outposts of the conquerors, the lonely presidios and missions of the padres, the later American forts. These trading centers were at once great fur capitals, important trail junctions, huge fairs, stupendous market places. Their history is not only the history of the fur trade, it is the history of the immediate future of the whole West.

There were three of these trading centers. One was Bent's Fort on the Arkansas—still on the plains, at the foot of the Rockies in southeastern Colorado. The second was Taos on the upper Rio Grande—in the Sangre de Cristos of northern New Mexico, the first settlement struck in the mountains and the last northern outpost of the crumbling Spanish empire. The third was the greatest of all, the one common rendezvous. Neither a fort nor a settlement, it lay almost in the center of the whole vast fur country, in the heart of the Colorado River basin: the "General Rendezvous" at Brown's Hole in the Green River Valley.

Its recorded beginning was an advertisement in the *Missouri Republican* of March 20, 1822:

THE TRAPPERS

To Enterprising Young Men:
The subscriber wishes to engage one hundred young men to ascend the Missouri River to its source, there to be employed for one, two or three years. For particulars enquire of Major Andrew Henry, near the lead mines in the County of Washington, who will ascend with and command the party; or of the subscriber near St. Louis.

(Signed) William H. Ashley.

Ashley was a Virginian who had migrated to Missouri, become a general in the State Militia and then lieutenant governor. Suddenly his luck had turned. He was defeated in a campaign for the governorship and went bankrupt to the tune of a hundred thousand dollars. At this point he met a veteran trapper and resolved to recoup his fortunes by entering the fur trade.

He could have picked no better partner. In 1808 Andrew Henry had discovered South Pass through the mountains of Wyoming and led a party into the valley of Green River. The river was still known by its Crow name, "Seedskeedee," rather than by Padre de Posada's "San Buenaventura." Here was a wonderful central valley where thousands of Indians gathered, and trappers came with their loads of pelts. Certainly this was the place to buy beaver and to stake out new untrapped territory.

The first hundred men started out under Henry. On their heels were a second eager hundred under Ashley. In unwieldy keelboats they followed Lewis and Clark's route up the Missouri. Arickaras and Sioux attacked so savagely that Ashley had to call back Henry for help, and then both for soldiers from the nearest fort. But when those left did get there with Henry they included some of the most significant men of the time: Jim Bridger, William Sublette, Hugh Glass, Jedediah Smith, Etienne Provot, Seth Grant, and Thomas Fitzpatrick.

Annually joined at Brown's Hole by Kit Carson, John Colter, Ezekiel Williams, William Wolfskill, Jim Baker, Baptiste Gervais, Louis Vasquez, David Jackson, Ewing Young and Bill Williams, these were the men whose wanderings covered half a continent, whose lives spanned two civilizations, and who were the prophets of the future.

Jim Bridger: the "Daniel Boone of the Rockies," perhaps the greatest mountain man of all. Kit Carson, acknowledged the best scout ever produced by the West. John Colter, who in 1807 reached the headwaters of the Green and discovered the Yellowstone region. Ezekiel Williams, already trapping alone in the Big Horn Basin. Jim Baker, who later led "Baker's Rush" into the wildest mountains of southwestern Colorado after diamonds reputed big as fists. Thomas Fitzpatrick, called by the Indians "Broken Hand," chief of the mountainmen, who led the first gathering on the Green. Jim Beckworth, a mulatto, made war chief of the Crows. Old Bill Williams, fantastic figure for whom a town, mountain and river are named . . . Each of these men requires a biography of his own. And none of them can be confined in a book of paper.

Ashley meanwhile had returned to St. Louis. Sometime later he met the fanciful mulatto, James P. Beckwourth, Beckworth, or Beckwith as he was called among the mountainmen. In November, 1824, they started out with twenty-four others to join Henry's group. By April 1, 1825, they had followed up the north fork of the Platte and crossed the Continental Divide. Here a band of Crows drove off their horses. After eighteen days of carrying the packs on their own backs, they came down to the Green. From "the quantity of wood cut along its banks and other appearances," Ashley noted that "it must have contained a great number of beaver" which he surmised must have been trapped some years before by trappers in the service of the Northwest Company.

He dispatched six of his men north to the source of the river, seven to a mountain in the distance, and six south to hunt Henry's band. With those remaining, Ashley made preparations to descend the river with the heavy packs "to some eligible point about 100 miles below, there deposit my merchandise, and make such marks as would designate it as a place of General Rendezvous for the men in my service."

Making frames for two mackinaw boats and covering them with buffalo hides, they set out on April 21st.

A week later they reached the base of a "lofty rugged mountain" where a creek (Henry's Fork) entered from the west. Camped that night, Ashley duly marked it as his "General Rendezvous." Where the river went he did not know. It seemed to disappear into the base of the mountain in a whirlpool that fanciful Beckworth aptly named "The Suck." Next morning they pushed off again.

At the first rapids, a fall of some ten feet which required a laborious portage, Ashley inscribed on a huge boulder his name and the year 1825. For ninety years it remained visible to record the first entrance of a white man in this wild and sinister river cañon. The red rock walls kept rising almost perpendicularly from the water to an immense height. The current increased. There was no getting out.

Running rapids, going six days without food, and in despair of ever escaping the cañon, Beckworth avers they were ready to select one of the party to eat. Then suddenly the mountain walls drew back, the river widened, and they shot out into beautiful Brown's Hole. Ten miles below was a great camping ground where thousands of Indians had wintered. Between Red Cañon and Lodore Cañon, just below the Wyoming line where Colorado and Utah touch, this was the one great rendezvous of all the mountainmen.

But by now the river was in Ashley's blood. The men

knew it. The gloom on their faces equaled that of the cañon before them, whose "massy walls excluded us from the rays of heaven and presented a surface as impassable as their body was impregnable." Nevertheless, they continued on downriver. Through Lodore Cañon, Hell's Half Mile, Echo Park, Whirlpool Cañon and Split Mountain Cañon; past the mouth of the Yampa, then the Uintah. Here Ashley finally abandoned the boats. Buying horses from a band of Utes, the party returned north, ascending the Uintah "to its extreme sources in the Uintah Mountains" of northeast Utah before joining Henry's men.

Ashley soon returned east, selling his fur interest to Sublette, Jackson and Jedediah Smith, who in turn sold out to Fitzpatrick, Bridger and others. Nevertheless, he had done a good year's work. He was the first man to navigate the main tributary of the Colorado and some of its worst rapids —and these in a couple of bullboats whacked out of a cottonwood and a buffalo in only two days. And he had established the one common rendezvous for all the mountainmen.

Here whole tribes of Indians came to trade, lining the river valley for miles with their smoke-gray tepees. Representatives of the many fur companies came driving long trains of pack mules loaded with whisky, powder and lead, with salt, sugar, coffee and staple trade goods, mirrors, beads and bright cloth. Here too once a year drifted all the solitary trappers from their far hidden valleys and nameless streams.

What a picture it must have been! Perhaps only one man saw and recorded it with the eye of an artist: Alfred J. Miller, a young painter from New Orleans.

Hired by Sir William Drummond Stewart, an eccentric Scotsman who was the Baronet of Grandtully, hereditary owner of Birnam Wood, he traveled with the American Fur Company's supply caravan from Independence, Missouri, to the Rendezvous of 1837. Mile after mile he recorded the scenes about him. Bearded teamsters constructing a bullboat to cross a river; swift Indian horsemen clinging to the shaggy fringe of a thundering buffalo herd; the caravan threading the limitless grassy plains; and finally the vast encampment at the foot of the jagged white peaks—in vivid water colors they all leap forth in timeless detail.

It is all one picture of a new fresh world gleaming with the pristine purity of its early morning, with all the wild earth's nobility, its savage barbarity and inexhaustible vigor. A vast and splendid panorama seen with the objective innocence of the eye of an artist. And caught in the supreme moment of fulfillment before it collapsed—just three years before the last rendezvous of 1840.

The value of these little known and but recently discovered water colors lies precisely in this panoramic scope and this wonderful freshness and purity which they evoke. The brush is mounted on wheels and never stops moving, it never digs into the body of the land. The depth, the distortion, the sublime exaggeration, the stark brutality, the mystic overtones, all are lacking. And it was precisely these qualities of inimical resistance which made the trappers what they were.

The yearly rendezvous was their moment of human contact. They leapt into it with bare fists and naked knives, kicking in the groin, gouging out eyes, biting off ears in the clinches, and stamping on fallen adversaries. Here Kit Carson fought his famous duel on horseback with a huge Frenchman, Bully Shunar. This was no panorama. It was a fantasy, a saturnalia. Everything swam in a cedar-red haze of raw corn liquor—the thousands of campfires and the Indians dancing around them all night, the incessant racing of horses against spotted Indian ponies, the gambling games carried on wherever a blanket could be thrown. High falsetto yells, the steady low beat of drums, shattered a silence unbroken for months. Above it all rang out the one secret voice that each heard separately, the multitudinous cry of all the lusts and hungers denied for months.

The trappers were no longer like beasts. They were like savages. They fought anybody, bet on anything, swilled whisky out of the kegtaps. Cheated by the traders, they traded fast and contemptuously, stripping themselves of bales of beaver and their several thousand dollars as they did their lusts—everything that had accumulated to hamper the strange, wild spirit of their unknown destiny.

The orgy over, they vanished again into the wilderness with the little needed to keep them going another year—or with nothing at all. Vanished again into the immense solitudes of unknown peaks, undiscovered cañons, the maze of unmapped streams. Where they went no man is certain, save that of all men who have ever lived none were given to span such horizons. Unknown rivers ran in their blood, fury made home in their hearts, and a vast loneliness that even the Rockies could never fill engulfed them forever. Their lives were like the winds that sweep down the cañons; you can hear their voice, but can find no footprints.

This is the Patties' epitaph.

Long before, a man named Pattie had followed Daniel Boone into the wilderness of Kentucky. His son, Sylvester, took up the trail into Missouri. For a while a sawmill on the bank of the Gasconade and a wife and children delayed him. Then with the death of his wife, Sylvester Pattie resumed the trail with his own oldest son. When he died in turn, this grandson, James Ohio, continued. For three generations the Patties were on the trail. It is a significant commentary on what we know of the last two.

The two Patties, son and grandson, came up the Arkansas, entering the mountains by way of Taos. Thence they cut across country southwest to the "Helay," discovering this (the Gila) to be a rich beaver stream. But just as they were ready to leave, the Indians stole their horses. Unable to carry their wealth, the trappers buried their bales of fur and walked back to Santa Fe. Here, clothes and moccasins worn out, they obtained supplies and horses to ride back after their cache of furs. The Indians had stolen them.

In despair, Sylvester engaged in a second business venture that promised better than his Missouri sawmill. He leased the ancient Santa Rita mines from their Spanish owners after beating off the Apaches who had prevented them from working. James Ohio kept going:

Down the Gila to the "Red"—the Colorado. Up the Colorado to the mouth of Bill Williams Fork, passing through the country of the "Mohawas" (the Mojaves). On up the Colorado, the first trip along the rim of the Grand Cañon, the first white man to reach Black Cañon. Still up the Colorado, having a brush with Shoshones. Over the Continental Divide to the Platte, and on north to the Yellowstone, arriving in May, 1826.

By summer he was back again in Santa Fe, this time with pack horses loaded with beaver. But the Spanish governor confiscated his furs on the pretext that they had been trapped without a proper license. "Excessively provoked," Pattie continue on south through the "Coco-mare-copper" tribe (the Coco-Maricopas) into Chihuahua, northern Mexico. Finally he drifted back to visit his father at the Santa Rita mines.

The elder Pattie had been having his own troubles. Long bored with the mines, he had just discovered that one of his trusted Spanish associates had run off with all the proceeds of their work. It was now the fall of 1827. Once more father and son struck out on a trapping expedition. This time they secured permission from the Spanish governor of New Mexico.

By December they had ridden down to the junction of the Gila and Colorado. Here the wily "Umeas" made off with their horses, and fearful of retribution even abandoned their camp. James Ohio had already been up the Colorado. Sylvester believed that there was a Spanish settlement at its mouth. Wherefore they set to work hacking out eight canoes united in pairs to carry their huge catch of beaver downriver. In this flotilla they started down through the lush delta region of the lower Colorado.

How wild it was and teeming with life! Flocks of waterfowl vast as clouds. Deer, foxes, panthers and wildcats. "Likewise an animal not unlike the African leopard," which they had not seen before. And beaver! Sixty a night they caught. For seven weeks they floated slowly along, trapping as they went, building more canoes to carry their furs. But now they kept constant watch for Cocopahs, Yumas and "Pipis." The Indians became bolder. They stood on shore waving their bows and arrows—a warning the Patties did not understand.

That night as they slept it happened: the rare phenomenon which distinguishes the Colorado among the rivers of the world: "the rush of the tide coming in from the sea, in conflict with the current of the river. At the point of conflict rose a high ridge of water, over which came the sea current combing down like the water over a milldam. . . . In twenty minutes the place where we lay asleep . . . was three feet under water."

By their own abnormal quick-wittedness the Patties saved their lives. Caching what furs they could recover, the two men set off on foot. But instead of the Spanish settlement they expected to find at the mouth of the river there was nothing but a barren, uninhabited seacoast.

Turning west and north they struck out across the alkali flats, deserts and mountains to the missions along the Pacific coast of California. At San Diego the Spanish authorities seized them and confined them in separate cells. Here Sylvester Pattie died, refused even a last visit from his son, James Ohio.

THE TRAPPERS

Eventually, after the prisoners' cache of furs had been found and appropriated, James Ohio was released, broken in health and penniless. Seeking indemnity from higher authorities, he made his way to Mexico City, where he was again disappointed. Finally he reached New Orleans by boat, and thence Cincinnati, Ohio.

In the whole history of the West there is nothing to match Pattie's wanderings. Not only did he make the first traverse of the whole Rocky Mountain region and the Colorado River from source to mouth, but one that has never since been surpassed.

Yet apparent failure, like an evil star, still hung above him. In Cincinnati he fell into the hands of a certain Reverend Timothy Flint who was then editing the *Western Review*. As a result, the *Personal Narrative of James O. Pattie* was published in 1831 by John H. Wood, with an Editor's Preface by Flint affirming that the narrative was written entirely and solely by Pattie. It is an amazing document. Historians, topographers, ethnologists, scholars and "Americana-ists" have protested it for over a century.

Much ado has been made over Pattie's excessively romantic interludes with a Spanish girl whom he saved from Comanches near Santa Fe. His continuous Indian and grizzly bear fights and other "blood and thunder clap-trap would have curled the covers of a yellow-backed thriller" of the time—with which it is classed. Of more importance is the fact that from his hazy topographical descriptions it is impossible to trace his exact route anywhere. "A clear, sane story of that one ride, small part though it was of Pattie's sum total of amazing wanderings, would have advanced the world's knowledge of the River of Mystery more than a quarter of a century." As it was, the book is still regarded by "practical realists" as worthless.

Consensus is divided on what happened. Either the Reverend Mr. Flint, following the same custom as editors today, edited out the valuable passages and interpolated the claptrap presumably more interesting to the readers; or, as more generally held, Pattie was gifted with little powers of observation.

A wonderful judgment! It requires its own jury.

For other mountainmen made notable journeys. William Wolfskill from Santa Fe to California by way of Gunnison Valley and down the Virgin River; Ewing Young from Taos to San Gabriel by the southern route; Dick Wootton from Green River to the Columbia and the Pacific; Kit Carson, Jim Bridger, Bill Williams and many others.

A few of them, like Pattie, wrote or dictated narratives and diaries of their lives in the mountains, as the *Life and Adventures of James P. Beckworth,* the fanciful mulatto; Ashley and Wootton; even George Frederick Ruxton, for whom the first little mountain creek I dabbled in as a boy was named.

A wealthy, highly educated young Englishman recalled to his ancestral estate upon the death of his father, Ruxton could not escape the inexpressible fascination of the mountains. He had to return. His narrative too is keyed to the same pitch of almost unbelievable exaggeration. Indeed, they all adhere to the same pattern as the famous tall tales of old Jim Bridger, who would casually tell of fleeing for his life from a grizzly, while sprinkling the ground in front of him with the astringent water of Alum Creek which shrank the distance to safety. Like their authors, these narratives are a peculiar product of the American wilderness.

Strange narratives indeed, compared to the diaries of the padres. Every trail that Escalante, Kino and Garces broke is described with such meticulous accuracy that they are imprinted on our maps today. Yet those of the trappers— who knew the country as the padres never did—are lost in a maze of unrecognizable cañons, peaks and rivers.

What is this strange paradox?

The truth lies in the men themselves. Both were suddenly confronted with that great psychical entity which was the heart of a new continent. The Spanish priests were meticulous, rational men, completely self-enclosed. They observed and recorded its physical outlines with utter detachment. But they did not touch it any more than they touched the expression of it in the Indian. And so they funked it; and in funking it, they were refuted by the land itself without ever understanding why.

The American trappers—conglomerate in origin as they were—were a different breed. Led west by the same strange instinct, they too were confronted not only by that high wall of the Rockies, but by the spirit-of-place of a new continent. The impact was tremendous. It shattered them completely, stripping them of all but the blind, stubborn will to resist.

And so in them we have the great drama of a new race and a new psyche derived from the soil of Europe combating the tremendous force of the American earth.

On the surface, the physical struggle to surmount the lofty peaks, to ford the turbulent rivers, to cross waterless deserts. To live like beasts, crisscrossing a thousand miles between campfires with the sure instinct of birds migrating with the seasons. What need had they of the detached rational observance of the padres? A prickling of the scalp, a tingle up the spine or less, were their compass and barometer, warning them of danger, change of weather, and the slow revolving directions. Wholly, intuitively, they gave themselves up to the forces surrounding them.

And at the same time, deep inside, the psychical combat. The great bulks of the mountains heaved up inside them, as around them. Space ate into them. The vast horizons stretched the last thin membranes that separated them from the body of the land. Rivers ran in their blood. And finally loneliness engulfed them, more vacuous than the depths of gorges below them, than the spaces between the stars above. They were stripped then to the eternal nakedness of their souls. There was no turning back. In their deep subconscious they knew it.

So they fought against everything that would have held them from their strange, unknown destiny. Against the lusts and comforts of civilization, against companionship, against home. Above all, they fought against women—against the physical contact that would have lessened the tension of their bodies, against the tenderness that would have alleviated the fury in their hearts.

And they continued pitting themselves against the spirit of the land long after they had learned to survive under its physical manifestations. Against the fabulous, weird and ever-changing phantasmagoria of the landscape with its perpetual mirages, its enigmatic and inimical face, the voice of its thundering, its whispering silence . . . This is one of the great truths in their narratives; they take on the dreamy anonymity of Currier and Ives prints, the agonizing stark nakedness of detail of Doré's pictures of the Inferno.

Still the tension kept mounting within them. And so they struck at everything. Not as the Conquerors, with a true and innocent blood lust, a simple will to dominate. But with a cold dispassion that frightened even the Indians. These they killed as they killed animals, on sight, as if this were the only natural outcome of a meeting. They seemed insensible to life—and most of all their own. And they died as they lived, unknown and alone. "Rubbed out," they would say of one another, missing at Brown's Hole. For it was not the arrow that killed him or the claws of a grizzly. It was the oppressive forces of solitude that finally crushed him into the hollow where he fell.

But little by little, as the trappers won the outward fight, they lost the inner. They began to get a glimmering of the Indians' conception of the earth as a living entity.

Many like Old Rube, as he was known, followed the Indian tradition of blowing smoke to the sacred directions. But the strangest of all was Bill Williams. Little was ever known about him save that when young he had been a circuit-riding Baptist preacher. He would appear like a ghost: a tall, gaunt old man with long hair and beard, astride a shaggy Indian pony. After a meal, perhaps a day or two with a party, he would disappear as silently and completely. Where he went no trapper knew; it was said he had a supernatural faculty for finding his way about and for detecting the presence of Indians. Certainly he spoke the dialects of many tribes, knew a great deal about their ceremonialism, and was an adept in their faith.

The whites reported that he believed his spirit would incarnate after death in the body of a bull elk, and they reassured him that they would not shoot him. Distorted as these white tales are, and probably Bill Williams's own incomplete understanding of the true Indian belief, they indicate how far he had traveled toward their concept that all matter has a spiritual essence as well as a material composition, that even mountains have a spirit form as well as a physical one. Fearful of the guardian spirit of one mountain particularly, old Bill refused to divulge its location even though he had found gold nuggets in a stream running from it.

Other trappers began to make peaceful contact with the tribes and consult their medicine men. Some had been accustomed to pick up a squaw to dog their heels for a thousand miles; she was fat and warm to sleep with and, better, she was handy at curing pelts. This was different. The trappers began to grow wholly Indian in spirit and feeling.

"Renegades," they were incorrectly called. But there were few, if any, real renegades. Perhaps three-fourths of all the trappers were killed by Indians—even at last Bill

Williams himself. Outwardly they remained aloof and solitary, white as always. But inwardly, becoming pure Indian, they began to establish the same fundamental relationship with the land.

In forty years, then, it had been done. They had completely cleaned every watercourse of beaver. There was not an Indian tribe which did not fear them, to whose last secret strongholds the trappers could not lead soldiers to annihilate them. They had opened up the last and greatest wilderness of America, the hinterland of the continent, to an influx of settlers that henceforth never ceased.

And yet they had lost as the Spaniards in this region never had, nor as the French and English east of the Missouri. Wilderness America had put its stamp upon them forever. They were never to outlive it. This is a tremendous psychical fact. Here, out of their aprocryphal conflict, was created a new breed. Men European on the outside and Indian inside, men neither wholly white nor wholly red.

For these were the first nonindigenous men to turn their backs wholly and forever upon their former homeland, to meet in mortal struggle the invisible forces of the land, and to accept its terms. And so they were the first of us, American "Westerners," to reflect, magnified and distorted, the signs of our subjection. Not until we understand the mountain men can we understand ourselves.

CHAPTER 4

The Settlers

No MAN is alone and isolated from his time. Even a solitary trapper crouched beside his campfire in the wilderness of the Colorado is susceptible to the impact of distant change. In destroying the consciousness of his own past, prophetic of the future, he looms as a symbol of what was happening the world over.

Across all Europe—in France, Austria, Germany, Hungary, Italy—geysers of unrest broke out. In an unparalleled outpouring of human emotion the tide swept over Europe, and kings ran before it in terror. All of South and Central America rose in revolt against their Spanish masters, establishing their independence. In North America, Mexico broke free from Spain and then the Republic of Texas from Mexico. The United States, declaring war against Mexico, took most of the Colorado River basin including what was to become Nevada, Utah, California and most of Arizona, New Mexico, Colorado and Wyoming.

What was happening? No one knew. History is not made from facts, figures, dates. History is made by living people. And people are swayed by mysterious, unfathomable impulses. Like plants and animals they react to unknown stimuli in the air and in the earth. Only when they have

forgotten what moved them, if they ever knew, or are dead
and mounted in the glass case which holds the mummies of
the past, are the labels tied on to them which are facts,
figures, dates—meaningless tabs on dead symbols.

1848 is such a tab, like 1933. Something then from
soundless depths gushed up into visible space.

Suddenly released, vast populations swarmed westward.
Happy, successful, contented, integrated people never up-
root themselves from their homeland. But for years, generations, a century, they continue to endure the conditions that
oppress them. Then suddenly they break free. Each individual perhaps has his personal reasons—political, religious, economic. They are all rational excuses. For not until the people
are under way do they realize they are taking part in a vast
revolutionary movement, a mass migration. They do not stop
to question why it is that now, at just this moment of time,
all of them at once have been impelled to simultaneous
action.

So from Europe they swarmed across the Atlantic. And
in America they broke suddenly across the brown Missouri.

The impetus that hurled them forward, across the tortuous Great Plains to the Rockies and beyond, was not, as
is commonly believed, the discovery of gold in California.
By 1849 the trails they followed were already broken and
trodden deep by great wooden wheels whose imprints still
show in soft limestone at the old crossings. There were three
of them, and a reason for each is now properly posted. The
Santa Fe Trail, established for the economic purpose of
wresting the trade from Chihuahua to the south. The Oregon
Trail, for the colonization of the Northwest. And the Mormon Trail, broken toward religious freedom.

Ahead of them was the high empty hinterland of the
continent just opened up by the trappers—the quarter million square mile basin of the Colorado. It was a vast vacuum

suddenly unlocked. The people had been waiting; they had been uprooted, were filled with a strange restlessness, beset with queer longings. And when the moment came they knew it. They knew it as birds know the moment for migra-

tion, as air rushes into a vacuum, as water overflows a broken dam.

Dollars, religious principles, homes—behind or ahead—meant nothing. Any trail would serve. It was a moment in the history of human migration that has few parallels.

The huge Conestogas rumbled along steadily but slowly, three or four miles a day. Only time had been speeded up. In just twenty years the resistless wave had washed across plains and mountains alike and had begun filling up the valleys, the meadows, the cañons, the whole vast basin of the great red river of the West.

It is a kaleidoscope of incessant movement. Seething clouds of dust from turning wheels obscure it, and the thunderous clouds of wildfowl rising from the watercourses never to return. The acrid smell of gunpowder and smoke fills the air. There is the crack of bullwhips over straining oxen; the scream of a mare with an arrow tearing through her lungs. Against the crash of forests falling under the ax, a shot rings out clear and sharp in a remote clearing. Men's voices are lost in the din of an empire being hammered out of the wilderness on the anvil of swift change. Men's principles are sacrificed to the birth of a monstrous and fantastic legend, the American Dream. This was no time for sermons. A six-gun was law. A lucky gold strike was fame and fortune. And there was always a bigger one, greener grass, more freedom, just over the next range.

It is a cinema of America on the march of conquest. But not a fantasy. We can reach out and touch it. That is its great reality. Time, change and movement—the only facts—have not only thrust it before our eyes but within reach of our hands.

Those wheel tracks end in the grave under the big pine. The wheels themselves are still sound where they lie, though cracked and warped. We know those tired voices, the touch of those wrinkled hands. They are the voices of our fathers and our fathers' fathers, they are the hands of our mothers. It is all ours, we know it all—the dust, the blood, the shame, the triumph, for we are the children of their creation.

But only now as the dust settles and the din fades, the real pattern begins to emerge. Oratory and reality separate, courage and fear mingle, the skeleton truth wears through the scarlet cloak. The pioneers, heroic demigods, become men.

But above all, they were immigrants. New people in a strange land, confronted by the force of alien natural laws inherent in the landscape itself. That is what they all faced. Their reaction to it and their growing perception of it is their only true history and ours.

The Explorers

The one link between the new land and the immigrants were those John the Baptists of the American wilderness, the trappers. Their trapping days over, they became guides for the long serpentine wagon trains. Many became professional hunters, as those hired at Bent's Fort, killing thirty or forty buffalo a day to supply the passing immigrants. Others became Indian traders. In 1843 Jim Bridger established a trading post on the Green River at Black's Fork. It became the junction of the Oregon Trail and the Mormon Trail that here split off from it. Louis Vasquez built a post on the South Platte for trading with the Cheyennes and Arapahoes. Sublette built another, the first permanent post in Wyoming, later Fort Laramie. Such posts became the government forts whose garrisons protected the new settlers. Other trappers, like Kit Carson, became Indian scouts leading the soldiers to the last secret strongholds of the tribes. Still others, like Jim Baker, led the first rushes of the gold prospectors. For all who followed, they were there waiting. Even at last for the railroad surveyors.

First instinct to point the way, then reason to substantiate it.

The official explorers, unlike the trappers, knew why they came. Unimaginative, orderly, bound by a sense of duty, they did not come to feel, to experience, to apperceive. They came to *know*.

Perhaps there was need for a little reason at least in the whole westward movement; if so, they were the only ones who supplied it. In 1803 Thomas Jefferson with blind instinct had bought the whole area pig-in-a-poke in the first place. Three years later a young lieutenant still in his early twenties, Zebulon M. Pike, was dispatched to see just what was included in this Louisiana Purchase. On November 23, 1806, he had crossed the plains with a few men, come up the Arkansas and seen the tall white peak they had believed to be a cloud hanging on the horizon. After naming it Pikes Peak and trying to scale it, he proceeded westward across the Sangre de Cristos and along the Rio Grande, which he mistook for the Red River. Here Pike was arrested by Spaniards for being on foreign soil and taken to Mexico.

Meanwhile an ugly rumor had grown up involving him. Aaron Burr was accused of conspiring to establish a new and separate empire in the western wilderness, and young Pike was believed to have been his secret agent sent to explore the region. The facts were never proved. Burr was later tried for treason and disgraced. Pike, released by the Mexican authorities, returned and was killed in battle during the War of 1812. Nevertheless, the incident was prophetic.

For fourteen years the two governments wrangled over the paper boundary of the Louisiana Purchase. They finally agreed that the southwest boundary should run westward along the Arkansas into the mountains, and thence northward along the Continental Divide. Then, in 1820, Major Stephen H. Long was sent with a party to explore the actual region about the boundary. From him derives the name of Colorado's highest peak whose melting snows drain into the great red river.

Not until twenty-two years later did the most publicized of all the official explorers arrive: John C. Frémont, the "Pathfinder." By then the trappers had trudged up every creek and cañon, spotted every peak. The country was full of trails—if a man could read the signs. It took the Pathfinder eleven years to find them, and with a trapper as his guide: Kit Carson. Altogether, between 1842 and 1853, Frémont led five expeditions into the Rocky Mountain region and across the Colorado basin along Escalante's Trail and the old Spanish Trail.

It was also Kit Carson who guided Colonel Stephen W. Kearny's Army of the West into New Mexico during the

Mexican War. The invasion was successful. On August 23, 1846, the Americans entered Sante Fe without firing a shot, were cordially received and given a salute of thirteen guns. The year before, Kearny had led an exploring expedition to Fort Laramie, Wyoming, and up the Sweetwater. Still later he led an expedition across Arizona to California.

By this time the population was streaming westward—more than sixty thousand along this route through Arizona alone. The government still wondered what the country was that drew them. So Captain L. Sitgraves in 1851 made the trip from Santa Fe to San Diego, sending back in his report, a Senate document, drawings of landscapes, birds, animals, snakes and plants. He was followed by Lieutenant A. W. Whipple in 1853 and Lieutenant E. F. Beale in 1857. These explorers, says the humorless Lummis, "gave us scientific assurance that we have here a desert as absolute as the Sahara"! Evidently the skulls and skeletons of those who had perished along the route had offered little proof.

But the explorers had duly explored the country and filled their reports with facts and figures which would stand until corrected by subsequent explorers. And Frémont had advertised it as well as himself.

Hence by 1849, when the California gold rush began, the three great trails westward had been already well established: the Santa Fe Trail to the south, the Oregon Trail to the north, and the Mormon Trail branching southward from it.

There were also three main routes across the interior of the Colorado River basin. One down the Gila River to its junction with the Colorado at Yuma, Pattie's familiar trail. Another by South Pass—discovered by the trapper Henry, Ashley's partner—around Salt Lake and down the Humboldt. And the third leading south from the Great Salt Lake by Mountain Meadows, and west by the Old Spanish Trail into Southern California.

The country was open and awaiting its first wave of settlers.

The Mormons

To repeat once again: it was a strange time in the history of men. Unintelligible whispers floated through the cornfields. Men saw queer visions as they slept at night. Their hearts were filled with ecstatic forebodings. It was a time when a new god might well be born—and many images of him were.

Everywhere thousands of people gathered to hear his prophets. Great religious conversions spread throughout the sparsely settled country. Western New York particularly became a focus of new movements. Under the influence of a Vermont farmer, William Miller, thousands of people climbed to the top of a hill and waited all day for the end of the world. Anticipating the Second Coming of Christ, the Shakers and Rochester Spiritualist Rappers arose. The Followers of Christ followed a prophet from Canada who forbade marriage, sanctioned promiscuous cohabitation, and had not changed clothes in seven years. In this same neighborhood a sixteen-year-old boy named James Collins Brewster began receiving revelations. Writing a book called *The Book of Esdras,* he began converting his neighbors.

It was not altogether strange, then, that another near-by farmer boy, Joseph Smith, on the night of September 21, 1823, received a visit from a messenger of God. The angel's name was Moroni. He described to the 18-year-old boy a book written upon gold plates, and two stones fastened to a breastplate, called the Urim and Thummim, which would enable him to translate it. Both the book and these seers were buried in the Hill of Cumorah near the dilapidated Smith farmhouse, between the towns of Palmyra and Manchester, New York.

After due time and instructions from Moroni, *The*

Book of Mormon was unearthed, translated and printed. A month later, on April 6, 1830, the Church of Jesus Christ of Latter-Day Saints was formally organized.

The Book of Mormon presumably is the history of the wanderings of three ancient tribes from the Tower of Babel which arrived in America about 600 B.C. By A.D. 420 the wicked Lamanites, from whom the American Indians were descended, had overcome the others, the Nephites. Mormon was the last of his race. Commanded by God to take care of the sacred records, he had buried them in the Hill of Cumorah, and designated his brother, Moroni, their heavenly custodian. Joseph Smith, in turn, was now divinely appointed to redeem the Promised Land for the spiritual Nephites.

The book was a failure. Mark Twain called it "chloroform in print." It was long, unimaginative, and almost unreadable for its errors in grammar, spelling and punctuation alone. Its style awkwardly aped that of the Bible. Not on one of its six hundred pages was there a new theological idea or the faintest reflection of spiritual insight.

Far from being accepted as divinely inspired, it was held to be a plagiarism from a manuscript written by a literary clergyman named Solomon Spaulding. Sidney Rigdon, a compositor in a Pittsburgh printing office, was accused of having stolen the contents of this *The Manuscript Found*. Knowing Latin and Greek, and having been a preacher, he soon became a Mormon leader and made a new translation of the Bible. Still later, when he was ousted from the Mormon Church, Rigdon established a rival "Church of Christ" based on the same plan. At any rate, *The Book of Mormon* didn't sell even when peddled from door to door, nor was it read through when given away.

Mormonism, the religion, was something else. It happened to be the spark that drew flame. Something in it appealed directly to the instincts of the uprooted, restless people. And Joseph Smith, the shrewd and uneducated farmer lad, something of a fake, possibly a paranoiac, became their Prophet.

But with converts came persecution. One throve on the other.

Within a year they were driven out of New York. Zion, the Promised Land, became Ohio, then Missouri. Soon the governor demanded that they be utterly exterminated or driven from that state too. With the massacre of eighteen members, the Saints fled to Illinois. Here at last the Temple rose; within five years the new Mormon town of Nauvoo became the largest in the state; the people began to prosper.

But by now Mormonism had become with slavery the great national issue, and of international concern.

Apostles under Brigham Young had gone to England, establishing branches in all large cities and a shipping agency to forward new converts. Joseph Smith could no longer conceal his toleration of secret polygamy or his own twenty-eight wives. He was mayor of Nauvoo, lieutenant general of the Mormon Legion, the Prophet of thousands of people, and in command of their national votes. One thing was left. He announced his candidacy for the Presidency of the United States.

Opposition against the Mormons suddenly crystallized; an insignificant incident released it. On June 24, 1844, Joseph Smith and his brother Hyrum were arrested and imprisoned in Carthage. Three evenings later a mob stormed the jail, murdering both Prophet and Patriarch. At Lima, near by, 175 Mormon houses were burned down. Saints everywhere fled to Nauvoo to throw up barricades. After battle and truce the Mormons agreed to leave Illinois, and were given two hours to evacuate Nauvoo.

Brigham Young assumed command. Unlike the visionary Prophet, he was a practical man, a statesman, a great leader.

On the morning of February 16, 1846, the first Saints

crossed the frozen Mississippi behind their plodding oxen. The great Exodus west had begun.

That summer they forded the brown Missouri between what are now the opposite towns of Council Bluffs, Iowa, and Omaha, Nebraska. Ahead of them lay the great wilderness of the West. Somewhere in it was Zion. Short of food, their horses and oxen worn thin with travel, they established Winter Quarters to grow crops and strengthen themselves for the march.

Next April it began: the first 148 people in 72 prairie schooners led by Brigham Young. A new Moses and his people seeking the Promised Land.

The huge Conestogas rumbled westward over the plains. For grease for the wheels they used the fat of marauding wolves. To mark the trail behind them they left heaps of buffalo skulls. They reached the Rocky Mountains, crawled up the pass.

Where they were going Brigham Young did not know. But from Frémont's explorations he had heard of a high mountain valley containing a strange salt lake. Jim Bridger confirmed it. The wagons creaked on. Brigham Young came down with mountain fever. Then suddenly, on July 24, 1847, he roused and looked out. An angel was standing on the hill ahead and pointing to the great mountain-encircled plain below, one of the most impressive sights in America. Zion!

Leaving most of the party to start a town on the spot, Brigham Young returned to Winter Quarters to lead the rest. The Promised Land had been found at last. The Mormon Trail was marked. Over it now swarmed an exodus that never ceased.

Among the later immigrants was a group of 1,300 poor converts shipped from Liverpool: Danes, Swedes, Welsh and English. Two-thirds of them were single women. Most of them could not understand English. They were proselytized from the poorer classes of large industrial cities and had never seen open country. Without covered wagons, even horses, they built themselves small rude handcarts in which to pull their babies, the sick and the lame, and their few belongings.

Thus they faced more than a thousand miles of unmapped plains and mountains. The Handcart Battalion. Theirs is one of the strangest and most courageous marches in all America. They were beset by wolves. When their food ran short, dysentery became an epidemic. Mountain fever attacked them. Cold overtook them; thirteen froze to death in one night. But God held back the one thing that might have destroyed them—the Indians.

Nor did anything stop those who followed. In the first ten years more than 76,000 emigrant Saints plodded over the Mormon Trail, 17,000 from Europe.

So here in the high mountain valley near the strange Great Salt Lake was established that which Aaron Burr had been accused of envisioning long before: a new and independent empire in the heart of the Colorado's wilderness basin—the Mormon Empire.

The State of Deseret, Land of the Honey Bee. That was the name under which it flourished. Comprising all of what is now Utah and Nevada, the greater part of Arizona, and slices of Idaho, Wyoming and Colorado, its tentacles reached into New Mexico, Oregon and California. It is one of the greatest parodoxes in the history of America.

In a nation founded upon and priding itself in its religious freedom, these people not only evoked the most violent persecution in its history, but in turn became the most fanatic persecutors of others. In the heart of a democratic republic they established a theocratic despotism. On the frontier of the West they maintained a polygamous family life almost Oriental in character—these people of northern races to whom the very idea of polygamy had been abhorrent for centuries. And finally, in the very period when rugged individualism was carving out the financial foundations of an imperialistic materialism, they founded the first successful socialistic state in the New World since the pre-Columbian civilizations.

No longer can we put the cart before the horse and say they did all this because they were Mormons. They were Mormons because of it.

Why? What were the causes that drove them to the utter extreme of all the concepts of their time and race?

Previous persecution is not the glib answer. Nor is divine guidance through the revelations of Brigham Young.

The trappers were the first solitary men to give themselves up to the forces of the wilderness heart of a new continent. The Mormons were the first integrated group. And both reacted the same way. The whole history of the Mormons, as a people, parallels that of the trappers as individuals.

To get *away*—to destroy thus the consciousness of the past. To immune themselves in this vast and lonely solitude against all the lusts and comforts of the civilization they had known, and to destroy all those who dared approach them; and, as the tension kept mounting within them, to strike at everything, to rend and claw themselves. This is their psychical truth as it was that of the trappers who preceded them.

The whole outer history of the Mormons is that of their isolation in space; their whole inner history is that of their isolation in spirit. From both springs their extraordinary conflict with the world of their time.

The first was challenged immediately. Gold had been discovered in California; wagon trains of gold seekers began invading the empire on their way west. The Mormons were immune to gold fever, and the few who left were warned not to come back. At the same time they were deathly afraid of the influx of "gentiles." Their persecution of these emigrants reached a bloody climax in the Mountain Meadows Massacre.

In 1857 a party of 136 emigrants were passing through Utah on their way from Arkansas to California. As usual, the Mormons were instructed to give them no aid, to sell them no provisions, and to show a negative hostility toward them. Early in September the wagon train stopped in the long mountain valley called Mountain Meadows. Here the emigrants were fired upon by Mormons dressed as Indians.

At the end of a four-day siege, John D. Lee, the Mormon leader, offered them safe-conduct to the settlement of Cedar City if they would surrender their arms. Accordingly, the emigrants marched off between two files of Mormon guards. Suddenly the latter opened fire, killing first the men, then wagonloads of women and the wounded. Of all the 136, only 17 children too young to remember more than their first names were killed. Their clothes were stripped off, their bodies mutilated and then piled in heaps for the wolves.

Lee reported immediately to Brigham Young, who in turn officially advised the United States Indian commissioner in Washington that it was Indians who had massacred the party.

Relatives finally traced the surviving children; the crime was solved; and nineteen years later, after two trials, John D. Lee was convicted and sentenced to be shot at the scene of the massacre. Before his execution, Lee made a full confession to retaliate against Brigham Young for having delivered him up as a sacrifice to save the name of the church. For all its ramifications, this most horrible crime of the Mormons stems directly from their fanatic desire for physical isolation.

Before it could be broken, the United States government was forced to send an army of 2,500 soldiers under General Albert Sidney Johnston to besiege Salt Lake City. When they arrived, the city was deserted. Thirty thousand inhabitants had loaded their wagons and started south, burning crops and grass behind them. A few hidden Mormons still left in Salt Lake City were instructed to set fire to the town at the first shot. The Mormons were determined to make "a Moscow of Utah and a Potter's Field of every cañon." Fortunately a peace was patched up; the inhabitants returned. And with the outbreak of the Civil War the army was recalled, General Johnston himself to die fighting a hero at Shiloh against the United States government. Altogether, this Utah Expedition, or "Buchanan's Blunder," cost the United States $15,000,000.

But thereafter the Mormon Empire as an isolated political entity was doomed. As the Territory of Utah it was whittled down on all sides: the western half becoming Nevada, that from the summits of the Rockies east given to Colorado, with a slice north to Wyoming, and a strip south to Arizona. Nevertheless, until 1870 the "ghost government" of the State of Deseret persisted as an ideal at once pitiful and inspiring, tragic and ridiculous. And not until 1896, a half century after its founding, was it accepted as the forty-fifth state of the Union.

Throughout all these changes, the people maintained an economic isolation. Tithing was upheld, whereby each man donated one-tenth of his whole crop, earnings or income to the church-state. To boycott gentile shops, every Mormon merchant kept a sign over his door with the symbol of the All-Seeing Eye. And finally there grew up the great cooperative corporation, the ZCMI—Zion's Cooperative Mercantile Institution.

With this physical isolation in space as a people there developed among them at the same time an increasing individual isolation in spirit.

The trappers, not to be held back by home and marriage, had refuted women almost entirely, becoming savage celibates. The Mormons went to the other extreme. Adopting polygamy, they isolated themselves behind dozens of wives.

That Brigham Young, for example, had twenty-seven wives and fifty-six children, and at the age of seventy was arrested for "lascivious cohabitation," has discolored the whole truth. Mormon polygamy was never the wholesale sex orgy that a generation of gentile enemies tried to prove. For the most part, the women were respected and well provided for; it was a social pattern made possible by their peculiar economy. The truth is that they could not bear the ideal of a close individual union—the electrical contact that both discharges a man's inward tension and fills his loneliness. Mormon polygamy, then, is a record of their terrifying mounting tension.

There was no spiritual affinity between a man and his many wives. How could there be? There was family loyalty, respect, a purely clannish relationship. But not a deep personal relationship. So little by little the spirituality, usually attained through the monogamous relationship, was diverted to a fanatical puritanism. Men must be kept righteous by any means. Their relationship with the other world was of primary importance.

This fanaticism climbed swiftly toward murder. To keep each other faithful to the church, men were beaten and flogged, then castrated and finally killed. Bill Hickman became known as the Destroying Angel for his prominence in carrying out these demands. As an executioner, he could scalp a man or beat him to death with the butt of his rifle with the calm assurance that he would receive eternal spiritual salvation.

From murder this fanaticism leapt to the last extreme —masochism. To sanctify it, Brigham Young brought forth the most horrible theory of Mormon theology—the doctrine of blood atonement for sins. Among these sins were apostasy, unfaithfulness on the part of a wife, and the shedding of innocent blood. "True love," he explained in a sermon, "was a love that would shed blood in order to insure for the loved one eternal salvation . . . will you love that man or woman well enough to shed their blood? That is what Jesus Christ meant."

The story is retold by Werner from the *Rocky Mountain Saints* of one of the wives of a Salt Lake Mormon who was unfaithful to him while he was on a mission in foreign lands. When he returned home she seated herself on his lap and confessed her sin, expressing the fear that for it she would be separated from him and their children in all eternity. The husband agreed; and as he returned her kisses, he cut her throat in order to win her spiritual salvation.

Under such terror of conscience, the Saints not only considered murder a righteous duty, but believed in self-slaughter. This blood atonement was a terrible doctrine. The world to come, with its promise of eternal salvation, blotted out the world in which they lived. It was the height of fanatical puritanism, the extreme of spiritual isolation. They could go no farther and survive. They had reached the last oubliette.

Thereafter their history is that of a gradual recall, both physical and psychical. Today, a few years later, their political and economic significance is purely local. As a sect Mormonism still exists minus polygamy, with a revised *Book of Mormon* and still maintaining tithing. The church claims a membership of 760,000 with branches throughout all the country, and missions in Canada, Mexico, Europe, Asia and Africa.

From its beginning in 1847 until its end in 1896, when Utah was admitted as one of the United States, the Mormon Empire is perhaps the greatest drama ever played on the wilderness stage.

But as drama it has obscured the meaning behind it. In seeing them as Mormons we have forgotten them as people. The first emigrant-settlers in this immense hinterland of

America. This is their tremendous, their only true significance.

For here nature forever domineers man—savage, untamed, demanding and devouring. Purple mountains tipped with white heap up on the horizons; yellow deserts, limitless and enduring, stretch into infinity. Space, limitless space, governs and shapes mere men in an invisible tyranny of mute terror. They cannot bear the sense of oneness, of aloneness, which the land inspires. So violence is their reaction— a violence inspired also by the stark majesty and savage brutality of the land. Violence and destruction, or an alliance with others pitted against nature. These are the only alternatives.

The Mormons, the first of us to settle here, reacted instinctively according to the first. We, rationally, have chosen the latter, conspiring with our machine civilization to conquer the land and its forces. Those to follow have yet to learn to accept, and not combat, and to find at last in its immensity their spiritual home.

CHAPTER 5

The Outcasts

The Prospectors

BRIGHAM YOUNG, warning his Saints against deserting Zion for the gold camps of California, said in a sermon: "Gold is good for nothing, only as men value it. It is no better than a piece of iron, a piece of limestone, or a piece of sandstone, and it is not half so good as the soil from which we raise our wheat and other necessaries of life."

This is a settler talking, the leader of the first emigrant-settlers into the basin of the great red river. The Saints obeyed their prophet. Utah is the story of a people rooted to the land, their Promised Land.

But even as the Prophet was thundering from his pulpit, gold was being discovered all around him: in the snowy mountains of Colorado to the east; in the deserts of Nevada to the west; and in the baked rock hills of Arizona and New Mexico to the south.

Gold! Bright, yellow, shining gold!

So on all sides of the weird vermilion cliffs of Zion swarmed a new type of men from across the plains—the gold seekers.

Unlike the Mormons, they knew why they came. They knew where they were bound for: that stream so full of gold sand a fellow can see it shinin' through the water!; that there mountain you could scrape a ton off by slidin' down it on a wooden drag! Looky here, this newspaper says gold's purty near everywhere—and Horace Greeley did.

So thousands rushed to this stream, that mountain. Whole towns built up on the first show of color. Then one night the wind brought the whisper of a bigger strike just over the range. By morning the town was deserted, and in a farther dark cañon mushroomed another one: Buckskin Joe, Tin Cup, St. Elmo, Bullfrog, Rhyolite, Elizabethtown, Piños Altos, Charleston.

That today is the wilderness basin of the whole Colorado. In it lie a hundred ghost towns, lost, lonely and forgotten in mountain and desert alike, populated only by wild burros and the forlorn specters of those whose extravagant hopes did not pan out; and a hundred living towns founded upon the fabulous luck of those who stubbed their toes upon treasures greater than their wildest imaginations had ever conjured. . . .

From the day young Pike first saw it hovering on the horizon like a white cloud, Pikes Peak was rumored to be rich in gold. James Purcell, whom Pike reported as the first American trapper in the region, had admitted finding gold there in 1807. In 1849 William Green Russell discovered gold in the Cache la Poudre on his way from Georgia to California. Then in 1857 George Simpson panned a few colors on Cherry Creek. This time the wind picked up the news.

Russell, who had straggled back to Georgia, heard about it and remembered the region. In February, 1858, he, his brothers and some friends of his Cherokee wife made up a party which reached the junction of Cherry Creek and the Platte in May.

In Lawrence, Kansas, a second party was organized by John Easter, the town butcher. This group was inspired by a Delaware named Fall Leaf, who reported having seen gold nuggets shining in a stream "two sleeps from Pikes Peak." Hastening across the plains that May, and camping for six weeks at the base of the peak without finding gold, they trudged north to join the first party. Here there arrived another party from Georgia led by John Gregory; and still another consisting of George Jackson, Tom Golden and an Indian named Black Hawk.

Gradually the men worked back into the mountains. On New Year's Day of 1859, at the junction of Chicago and Clear Creek, Jackson made the first big strike. From it mushroomed the gold camps of Golden, Jackson Bar, Black Hawk and Idaho Springs.

In May, Gregory struck it on Clear Creek, giving rise to the Gregory Diggings and Central City. Just above, and in the same month, Russell found gold in the gulch named for him. It was a district that before the end of the century was to produce $186,000,000 in gold.

The Pikes Peak Rush got under way. By summer more than 100,000 people were swarming across the plains in lumbering Conestogas with "Pikes Peak or Bust!" emblazoned upon their canvas. The California forty-niners derisively called these Colorado fifty-niners "Pikes Peak Skunks," and there was reason for their jealousy. It was a rush whose intensity has never been equaled.

For though a year later a thin stream of disconsolate and destitute gold seekers trickled back—their tattered wagon sheets now printed "Busted, by Gosh!"—nothing stopped the flow.

Within three years the few cabins on Cherry Creek and the Platte had built up into the city of Denver—a ripsnorting community huddling at the base of the mountains nearly 1,000 miles from rail ends and navigable water, and yet receiving 1,000 new arrivals daily. The region was organized as the territory of Jefferson, but admitted in 1861 by Congress under the old Spanish name given its red earth and red rivers—Colorado. Chosen as its first capital was the camp located at the base of the Pikes Peak, El Dorado, renamed Colorado City.

At the first Fourth of July celebrated in Denver, the national salute was fired with an extra shot each for Pikes Peak, William Green Russell and John Gregory.

It was the one memorial salute ever given these early discoverers of gold in the Rockies. Gregory—in whose gulch Central City is still mounted like a gem—spent his few thousands in Georgia and then returned to disappear in the mountains of Colorado, still searching gold. Green Russell, who also went back "home," returned to the great San Luis Valley hollowed in the Colorado Rockies and died among his wife's people, the Cherokees. Even Jackson, who left to serve gallantly in the Confederate Army, was drawn back to trudge the lonely trails through the wild San Juan and La Plata mountains. Never making a substantial strike, he settled on a ranch at Chipeta, Colorado, where he accidentally shot himself.

Pikes Peak still remained the beacon of those Shining Mountains of the Spaniards which "would be found to contain more riches than those of Indostan and Malabar, or the golden coast of Guinea, or the mines of Peru."

From here even farther into the great dark cañons pushed the prospectors. "Let's all tarry here and have fair play," agreed Joe Higginbottom, who always wore buckskin clothes, and his partners. So in that great upland meadow—South Park—there grew up the camps of Tarryall, Fair Play and Buckskin Joe.

They scrambled higher and higher up the great blue wall, founding Breckenridge on the Blue River. Mines were opened at Tin Cup, above St. Elmo, one of the weirdest ghost towns of the Rockies, and up one of the wildest and most inaccessible cañons.

Finally they reached the top of the world, two miles above the tidewaters of the Pacific to the west, the Gulf of Mexico to the south, and the Atlantic far to the east. Between patches of April snow four feet deep, Abe Lee, a discouraged forty-niner, was found staring into his pan. "Boys!" he shouted, jumping to his feet, "I've got it—the hull of Californy in the bottom of this here goddam pan!"

Thus California Gulch produced in two years over a million in gold. It was a mere grubstake toward the later discovery of silver; enough—it was boasted—to build a solid wall of silver four feet high and four feet thick across the whole state. Amid mines that poured forth as much as $118,500 in one day grew up Leadville, the Cloud City, "a silver city in a sea of silver." Yet its approach up the pass was called "The Highway of Frozen Death." Here was a single camp that produced the American phenomenon called H. A. W. Tabor, that built the Guggenheim fortune, and laid the foundations of the Marshall Field and the Flatiron Building in New York, and is one of the really great fables of America.

And now with a shout of triumph from the summit of the Continental Divide, the prospectors tumbled headlong into the great basin of the Colorado.

In the little streams of melting snow trickling off the fifty-one 14,000-foot peaks which feed the great river; in the headwaters of the Blue, the Eagle, Gunnison, Yampa and the Dolores, they found gold.

Kokomo, Robinson and Red Cliff sprang into being. Aspen exceeded for a time even Leadville in size. Here, out of the Smuggler, was taken the world's largest nugget, 2,060 pounds, 93 per cent pure silver. Down in the Sangre de Cristos, to beautiful Wet Mountain Valley, rushed men to create Rosita and then Silver Cliff which in its brief heyday followed Denver and Leadville as the third largest city in Colorado.

Jim Baker in 1860 had led the first rush into the mysterious San Juan Mountains of southwestern Colorado, getting as far as the headwaters of the Rio de las Animas, the River of Lost Souls. It was appropriately named. For, although rumors insisted that Baker's band picked garnets, rubies and diamonds off the ground, all the men perished trying to find their way back. From this Baker's Rush derived the expression "over the range," alluding to one who has died. Now, ten years later, gold was found in the region. Long files of pack burros snaked down the divide. In the King Solomon District, Nat Creede's obstreperous burro kicked loose a specimen of gold so rich that he could only mutter "Holy Moses!"—thus naming the mine that soon began producing $180,000 a month. Creede, Telluride, Ouray boomed into being, and then Silverton, where Thomas Walsh discovered the $25,000,000 Camp Bird.

As Utah is the story of the Mormons, so Colorado belongs to the prospectors; its state emblem still bears their pick and shovel.

Meanwhile, two miles below and nearly two thousand miles south of the source of the Colorado, gold had been found near its mouth. In the same year, 1859, that the Pikes Peak rush began, desert rats struck pay dirt twenty miles east of Yuma at Gila City and seventy miles north at La Paz. Two years later at Prescott a member of the Joseph Walker party, Jack Swilling, dug out $4,000 worth of gold with his jackknife. It was the beginning of Arizona's boom.

In 1864 Henry Wickenburg, an Austrian teamster, discovered the rich district named after him, and a mine that was to produce $80,000,000 in gold. This was followed by the great gold, silver, and copper strikes at Globe, Bisbee, Miami and Jerome; and later, Ed Schieffin's discovery in 1878 of the Tombstone district which produced nearly $40,000,000. Like Colorado, Arizona was admitted into the Union as a territory in 1863 because of its rich ore discoveries, despite the existence of a strong prosouthern sentiment and its recent occupancy by a Confederate force.

In New Mexico the prospectors followed the beaver streams of the trappers Jedediah Smith and old Bill Williams with equal success. Gold was discovered at Elizabethtown, and just north of Silver City at Piños Altos, and later at Blanding.

Far to the west, in the lowest spot on the continent, gold was found in Death Valley just as bright and yellow as that on the highest crest of the Rockies. Here, too, in a new strange wilderness of treeless desert, camps sprang into being, new names to ride the wind: Panamint, Darwin, Ballarat.

And lastly in the desert of Nevada, Jim Butler, searching for his burros in a dust storm, found gold in the ledge behind which they had taken shelter. He named the place Tonopah; and from his original investment of $25, it produced $150,000,000. Then years later gold was found a few miles south. The sagebrush district was first named Grandpah in derision of the Paiute names of Tonopah (hidden spring water), Weepah (tank water), and Ibapah (red water). "Grand-pah," they insisted, "will be the grandpappy of 'em all!" It was; in fact, as it grew older and more dignified, it took the proper name of Goldfield. A year later Bullfrog jumped on the map, named from its dark-green mottled ore; also Rhyolite, Rawhide and Eldorado.

Meanwhile Pikes Peak was having its last laugh. It was the magnet that had drawn the first rush of fifty-niners. But for thirty years swarms of prospectors had trudged across its slopes on their way to other fields, ignoring the claims of a mountain cowboy, Bob Womack, that there was gold in the peak itself.

Then on the night of July 3, 1891, a lonely prospector dreamed of a mine near a certain ledge of reddish granite he had passed. At daybreak he hurried back to the spot and filed his claim, naming it the Independence. From the twisting little stream below, the district took its name, the most fabulous in American mining history—Cripple Creek.

Over ten thousand feet high, on the side of Pikes Peak itself, it became overnight the richest single spot in the world. A treeless granite slope of scarcely six square miles, it supported at once 175 shipping mines, 11 gold reduction mills, and 9 samplers; three cities, six main camps, and the highest incorporated town in the world. It has produced over $450,000,000 in gold to date.

The Independence alone has produced $90,000,000. A mine so rich that its many veins were named like other single mines; an ore so pure a man could bring out a fortune in his pants pocket. Its underground chambers were locked with bank vault doors and it was called, in a report made by some of the world's most noted mining men who inspected it, "the most noteworthy gold mine of the Western Hemisphere, if not of the world."

W. S. Stratton, its discoverer, justly called the "Midas of the Rockies," was a man who matched his mine. At the age of twenty-six he contracted gold fever and gave up his business and family. For seventeen years he trudged the lonely trails behind a solitary burro before his first, only and incomparable strike. Eight years later he sold his mine for $10,000,000 cash. Yet confirmed to his strange search, still living the life of a recluse, he bought up one-fifth of the whole area of Cripple Creek and announced the fantastic project on which he was now to embark—the sinking of a shaft down into the core of Pikes Peak itself. Two years later, still pursuing this search, he died.

His is a story—both man and mine—that plagued me personally thirty years until I wrote it. We knew Stratton well. My grandfather bought out his early contracting business, and with the money Stratton went on his first wild-goose chase after gold. Later Grandfather employed him as a carpenter; and still later developed for him the great American Eagle mine whose portal is the highest in Cripple Creek. Stratton's solid old house was around the corner from ours. I still have on my shelf the old copy of Cornwall's *Manual of Blowpipe Analysis* that he carried in his pack. I first went down into the Independence slopes at the age of ten.

It is a story that has tantalized the imagination of the whole West as well. For in its gangue—in the matrix of its rich narrative ore—lies the psychological truth of all his contemporaries.

Stratton epitomizes the thousands of prospectors who preceded him, who have died unsung and often unfound in their last cañons. His was at once the luck of the few and the solitary privation of the many. He was a man who, even while he lived, was an enigma and an outcast, and whose legend today is compounded wholly of the peculiar mystery that envelops them all. To Stratton, as to all of them, wealth was inconsequential.

For, strangely enough, something paradoxical happened to all these men. Few of the great gold discoverers ever returned to the life of ease they had envisioned and achieved. Their fabulous strikes they sold for a song or squandered as quickly as possible so that they could hit the lonely trails again.

The men who made fame and fortune were not the discoverers. They were the men who bought the strikes, men cold, calculating and businesslike; men like Guggenheim in Leadville, Pullman in Central City, David Moffat, Newhouse. The sellers were like Gregory, Jackson and Russell, who returned to the mountains to die unknown. Wickenburg shot himself after discovering the greatest gold district in Arizona. Red-haired Ed Schiefflin, rich from the discovery of Tombstone, was found dead and destitute years later in a lonely cabin, still prospecting. Rische of the Little Pittsburgh was pulled from the gutter to become an old broken watchman in the Denver state capital. Even the great Tabor. . . . The list is endless.

Four centuries before, a group of men had come to the New World for the same thing—gold. The conquistadores

found it. And they loved it, the weight of it, the pale yellow glint of it, the feel of it. With it they bought power, founded great royal families, ruled whole provinces. Gold for the Spaniards was an end. The search for it, the land that held and gave it, never touched them.

But something strange happened to their successors. Gold to the American prospectors became a means, not an end. They themselves became a new breed among mankind. A solitary ragged fellow trudging alone beside his burro into the dark cañons; a solitary desert rat crouched beside his campfire on the illimitable sands: this is the immortal picture of those American knights who followed the gleam, the prospectors.

The prospectors were the blood brothers of the trappers. Each sought the same mystery and in the same streams. One called it beaver, the other called it gold. Both were empty names. Both were symbols. And when the searchers found them they knew it. It was their one weak moment of utter disillusionment, and their one great moment of fulfillment. Henceforth they knew the strange destiny to which they were inexorably confirmed. The worthless symbols—the bales of beaver and the nuggets of gold—they got rid of quickly, and continued the quest. This time it had no end, no name or shape for what they sought.

This is what made them seem queer to their fellows. This is what made them outcasts.

The Outlaws

> Scene: The board walk in front of the Last Chance Saloon.
> Spotlight: There he stands, suddenly frozen in a stiff crouch; hands wide open and dangling loosely before him, waist high; his cool gray eyes narrowed into slits in his handsome, somber face.
> Action, Sound: A breathless silence fraught with suspense. Then the draw! too swift for human eye to follow. The crashing shots of two blazing six-guns drawn from the hips and returned smoking to their holsters before the single report has died away. An instant later the sound of a horse's hoofs tearing hell-for-leather down the street, heading for the hills.
> Voice: "Gawd, boys! Johnny Two-Gun, the Killer!"
> Exit: America's best-loved hero, the Outlaw.
>> But don't worry, folks, he'll be back all right, all right. See him next Saturday night at your neigh-

borhood theater. Follow his killings meanwhile in your favorite pulp-paper westerns. And don't fret about him in his hideaway in the hills. He's immortal as hell. Them guns of his will be blazin' long after you and me, mere humans, wander over the range.

The Badman, the Two-Gun Man, the Six-Gun Man, the Squaw Man, the Killer, the Outlaw, as he is variously called, belongs wholly to the myth he has created.

But there have been plenty of men to contribute their lives to its building. From Arizona to Wyoming, from Colorado to California, the basin of the great red river has been his stamping ground. In it he has known many names: Wild Bill Hickok, Billy the Kid, Bat Masterson, Ben Thompson, Clay Allison, Jack the Ripper, Buckskin Frank Leslie, Luke Short, Doc Holliday, Curly Bill, Ringo, the Wolf of Washita, even the famous James brothers who were quietly run out of Leadville, to designate but a few.

They drifted to all the frontier towns in turn: to the mining camps of Colorado and Nevada, to the cowcamps of Arizona and New Mexico, to each of the rail ends across Utah. Some of them were professional gamblers. Others were confirmed cattle rustlers. Some were stagecoach robbers with an organization of town confederates, holdup men, even petty thieves and barflies. Still others were merely innocent ranchers, cowhands and young adventurers who loved liquor and noise on Saturday night.

But when it happened, the one was set distinctly apart. A few drinks in a crowded saloon, a quarrel, a drawn gun; thenceforth he was a marked man. Hiding in the hills, he came to town knowing that he would last just so long as he could live up to his sudden reputation of being quick on the trigger.

He became an outlaw. Let him travel hundreds of miles away, over snowy mountains or across a trackless desert, he always found that his reputation had traveled faster and farther. Also that it had grown. It not only encompassed that of the man he had succeeded, but several reputed others. It also took on the monstrous exaggeration of the land itself. There was no living or shooting it down. He could only practice fanning his gun faster. Eventually someone, somewhere, he knew, would be quicker.

Occasionally a few realistic citizens called to make him sheriff. The offer had its many advantages and few disadvantages. He would be more gunned after than before, but freer to shoot without taking chances and without having to run afterward. For the now confirmed gunman it was a happy choice. With a tin star he stood the chance of also wearing a halo of righteousness; without it, a noose. He usually took it, and this is the culmination of the myth, the American happy ending.

I have known two outlaws. One, a man; the other, a myth.

The man was one of my best friends. A half-Cherokee Indian, I shall simply call him Cherokee. At the age of twenty he reputedly killed a man in the Strip and finally hid out in a remote valley in the Sangre de Cristo Mountains. An old gambler took him in. The gambler, himself shady and retired, owned an old inn in a Mexican village, making his money dealing poker to visiting cattle buyers, surveyors and politicos sent up to buy the local vote. Cherokee did all the work, cutting wood, slaughtering beef, patching up the 70-year-old building.

The old gambler when I first met him was growing blind. But he had taught Cherokee how to deal. When he died, leaving his widow without help, Cherokee stayed on. He ran the inn for nothing but his board, and made his salary dealing poker at night.

He had been there twenty years when I moved in—the only permanent guest in the rattletrap old inn, and the tenth white (Anglo) in the village. Cherokee robbed me of a month's rent in the first game I sat in, forbade my ever participating in another, and thereafter looked after me like a child.

He was excessively high-strung, never set foot outside of town, and always slept with his belt and revolver under his pillow. One night while I was reading in bed, there was a sudden commotion in the road outside. Men's voices, the sound of horses' hoofs and a shot. The bullet tore through the window and embedded itself in the adobe wall beside my head, splattering glass and adobe so thick about me that the Indian girl employed as a servant had to come in and remake the bed.

But Cherokee, almost before the sound had died away, had jumped out of bed, run across the flagstone placita, and burst into my room with his drawn forty-five. A thin, dark little man with a tense face and that huge gun standing quietly in the light of the smoking lamp.

Mexican neighbors coming home drunk from the dance hall, I'm sure; pure accident. But enough to reveal Cherokee's state of mind even after twenty years.

The end was both tragic and humorous. I had gone to a town in Colorado for a few months when Cherokee called me one night from the railroad yards. I picked him up there and brought him home—wearing an old broadcloth suit with his gun belt under the coat and with the seat of his britches gone! He told me what had happened.

The day before had been court day. A group of visiting politicos had taken lunch at the inn. One of them had laughed at the Indian girl's biscuits. Cherokee, serving table, had immediately cracked him over the head with the full water pitcher. The victim happened to be the district attorney.

Cherokee fled and hired a Mexican to drive him over the Colorado state line. The man was drunk, ran into a barbed-wire fence and Cherokee landed on the other side with the seat of his pants torn out. However, he managed to catch a freight to the town where I lived. Next morning we read that the district attorney with a split skull had obtained his Oklahoma record, and offered $10,000 for his arrest.

 Till the news died down he stayed in my little tourist cabin on the outskirts of town. He almost drove me crazy. Hours at a time he would sit cracking his knuckles, kneading his hands with glycerine, and dealing cards like a magician. At last the letter came he was awaiting. A group of gamblers had offered him a job as dealer during the various summer fiestas in the Colorado mining towns—Ouray, Silverton, Durango.

 Two weeks later we packed up the Ford. I took him over the pass and then turned south to another town where I stayed that summer. Here I forwarded him mail under different aliases, occasionally contacted his friends and sent him money from them.

 Late one night he showed up, and I drove him back to the inn to collect his clothes and say good-bye to the only peaceful home he was ever to have. For twenty years he had known this high pass, these mountains; had dealt poker to every timber cruiser, surveyor and cattle buyer; had loved and cursed it as only a half-breed can. A young state policeman whom he had known as a boy slipped him out of town on his way to pick cranberries in Texas, and to become a vagrant after that. I never saw him again.

 The other outlaw, the myth, I knew indirectly. Of all places to stumble accidentally on his trail, it was in the city and on my mother's sofa.

 On it sat a weazened little old lady who each afternoon came in to rest as she peddled little bunches of withered flowers from door to door. Throughout the neighborhood she was familiarly known only as "Aunt Allie." Eighty years old and eighty pounds little, she was wholly delicious and perpetually cheerful as her dog "Twinkle." Aunt Allie was full of tall tales. Gradually and persistently she insisted that I write her "life story."

Reluctantly I agreed. She was the real thing—pure Americana, a yard long and a yard wide. For another, she was dependent for a cot and meals upon a family of distant relatives; her only spending money came from selling a few flowers, and she desperately needed the few dollars that the possible sale of her "story" might bring.

At first our talks went well. Aunt Allie was childish; the old years came back more clearly than yesterday. Wonderful years! She had been born in Florence, Nebraska, near the Mormon Winter Quarters. As a child she watched the Exodus, saw the Handcart Battalion pass by. Her chum was named Amelia—the same Amelia who later became the twenty-fifth and favorite wife of Brigham Young, and for whom he built the house in Salt Lake City that became known as "Amelia's Palace." Allie herself got married to a young teamster and made the overland trip by covered wagon to Prescott, Arizona, at the height of its boom.

It was, so far, a rare, firsthand pioneering saga that rang vividly true to the slightest detail.

Then suddenly something happened that changed it completely. We arrived on paper at Tombstone. I awoke to an astonishing fact. Her husband's name was Virgil. He had two brothers, Morgan and Wyatt. The famous "Fighting Earps" of Tombstone! Decidedly this was something!

But promptly her miraculous memory for detail failed here. And, curiously enough, she showed a sudden bitterness. Wyatt's wife had secured a man to write his biography, and for several years had bragged of the jackpot it had brought her. To Aunt Allie this was a sore point; she was jealous, and determined that Virgil should get his just due of fame. At the same time she would relate nothing but their family life in Tombstone—nothing of the fights and robberies, holdups and murders.

At this point her distant relatives came over with a paper for me to sign, guaranteeing them full monetary rights to her story in case of Aunt Allie's death. Then, unaccountably, Wyatt Earp's wife called in my absence with the threat that she would sue me should I publish anything on the subject of her deceased husband. The whole history, she maintained, had been already authentically covered in *Wyatt Earp, Frontier Marshal,* by Stuart N. Lake: Wyatt's purported autobiography.

Meanwhile I had been offered an old house near friends in Arizona to live in for two months. It was on the Mexican border between Nogales and Tombstone. Batching here, I began to dig into what already had become the Tombstone Travesty.

Research on Western Americana is not an armchair job. The year before, I had lived in a miner's cabin in Cripple Creek while chasing down the truth about Stratton. The trail wound between all the high frosty peaks of the Colorado Rockies, through a dozen mining camps, to over forty old-timers still alive who remembered him.

This time, in the same battered old Ford, I covered the Arizona deserts from end to end. I talked to nearly fifty old cattlemen, old-timers, judges and lawyers. The Arizona Historical Society gave me old letters, old affidavits, files of yellowed newspapers. Other letters and old court records came to light. All evidence whose existence had been denied by Lake.

Pieced together, they all settled for me this most famous and controversial subject of Arizona. Wyatt Earp had been little more than a tin-horn outlaw operating under the protection of a tin badge until he was run out of Arizona under President Arthur's threat to declare martial law in Arizona unless the Tombstone district was cleaned up.

Wyatt Earp, Frontier Marshal, his purported autobiography dictated to Stuart N. Lake, is the standard textbook adhered to by all movie and pulp-paper western writers. In it he is portrayed as the model frontiersman, a super-mixture of scout, plainsman, Indian fighter, buffalo hunter, marshal, and all-around Dead-Eye Dick—a Ned Buntline Special. Actually it is not an autobiography at all. It is the most assiduously concocted blood-and-thunder piece of fiction ever written about the West, and a disgraceful indictment of the thousands of true Arizona pioneers whose lives and written protests refute every discolored incident in it.

This is not the place to relate the historical discrepancies. Besides, I returned with a bigger and more human story. I had found what was behind Aunt Allie's silence.

Lake's book pictures Wyatt as a dashing single man in Tombstone boarding with his two brothers and generally protecting them with his prowess. Actually all three Earp brothers were married and went there together with their wives—in Wyatt's case, his second wife. Still in the county court is a record of a suit against Wyatt and his wife Mattie for recovery of money borrowed against their house.

Run out of Arizona, Wyatt abandoned her. Soon after, she was found dead of poison near Wilcox, possibly suicide. Earp went on to San Diego and made money in the real estate boom. Then he went to Alaska where U.S. Marshal Albert Lows slapped his face, took his gun away from him, and ran him out of Nome. Wyatt then returned to San Francisco as a sporting man. In this new role he became referee of the Fitzsimmons-Sharkey fight—which he undoubtedly threw to Sharkey on an alleged foul, having bet heavily on him. Here he married a wealthy San Francisco woman, Josephine Sarah Marcus, who would have nothing to do with the rest of the Earps—particularly Virgil and Allie. Lake's book, written at her order, therefore omits all the homely family life of the three brothers and every possible mention of Wyatt's second wife. It is, in short, a fictitious glorification of a man written to the order of his third wife.

With this I returned to Aunt Allie. The whole thing was plain. Morgan had been killed in Tombstone. A younger brother, Warren, a stage driver, was shot in a saloon fight. Virgil, maimed for life, eked out a 20-year existence in the mining camps of Nevada, followed faithfully by Aunt Allie. Only Wyatt, the tinhorn outlaw, remained to reap fame and fortune as America's most glorified frontiersman.

Little wonder that this destitute little woman of eighty had twitches of jealousy prompted by Wyatt's third wife's regular attendance at Tombstone's gala Pioneer Festivals. Aunt Allie alone had survived and lived it all, and was destined to die unheard. For, confronted by the facts, deathly afraid that Virgil's name might be "besmirked," she still refused to tell anything that might implicate him. It was a magnificent loyalty which endeared her to us forever: by it she showed what built Arizona—the integrity on which the whole West is built, and which today refutes Wyatt Earp's grandiloquent boasts.

Not until she dies will the book be presented to the Arizona Historical Society merely to complete its Pioneers' file as pure Americana.

For books, as books, are worthless. It is what they teach that gives them their only real value. And what we need is to understand the American Myth, the psychological truth of the men and the breeds that reflect the hidden truths in us today. The Tombstone Travesty is pertinent. Not only did its long investigation teach me what was the true outlaw, but it shows how the myth began. It shows who finally has created the outlaw, one of the truest figures in the American Parade—ourselves.

There's no use trying to act uppish about it. With all our vaunted rationalism and realism, we can't deny him. His appeal is singularly American and profound. More than any other, he embodies the secret loneliness in all our hearts, the uninhibited lust for violence, the naked fear, the relentless unrest. Like the trapper and the prospector, he was an outcast from his kind.

But here is the great difference. Unlike them, the outlaw was not an outcast of his own volition, and thus lacked the peculiar mystical strain which they knew as simply a strange and impelling wanderlust.

Both the trapper and the prospector was the new American self-driven forward to his own strange destiny.

The outlaw was the negative new American cornered by this destiny and fighting against it. The only true renegade. He embodies the stark-naked fear that lies at the bottom of all our hearts.

First of all, the instinctive fear of the ever-inimical Western American landscape with its limitless savage deserts, huge brooding mountains, and weirdly distorted cliffs —the fear of its spirit-of-place that made him feel a stranger and would not let him rest.

Fear, too, of its overwhelming immensities, the mere immeasurable space that dwarfed him to an infinitesimal moving speck. And lastly, the fear of its haunting timelessness which overemphasized the brief, dangerous and suddenly ended span of his own life.

Everything about the outlaw betrayed this unconscious, deep-rooted fear and inferiority. Lacking real strength, he had no gentleness. Lacking all but a desperate physical courage, he gave no odds and shot on sight. Without trust, he had the cold unsteady eyes which were forever wary of approaching strangers and friends alike. Even his face—the long western mold of fiction with deep cheek crescents—was an unemotional mask to match his taciturnity. Appearance and action, both added up to a complete and frozen inhibition.

A man wholly self-conscious, forever tense and unrelaxed and completely inhibited by his secret and unadmitted fear. Hence his boasts and the proud notches on his pearl-handled guns. A man with the temperament of a schoolboy bully who forever carried a chip on his shoulder to prove his courage—and who usually died with his boots on at the first instance his bluff was called.

He is the most to be pitied, for he suffered most. We understand this suffering. It is what makes him our favorite American.

CHAPTER 6

Its Travelers

During all this time when emigrants were swarming into the basin of the great red river and fording it to crawl still farther west, the Colorado itself emerged from mist and dust. It took on a legendary character of its own. One that embodied all the strange, terrible and sublime quality of the land it drained, and something of its own uniqueness.

Its legend persisted to my own day. We spoke of the Colorado as men spoke of the Congo and the Amazon. Its rocky red fastnesses were mysterious and unfathomable as the heart of darkest Africa. A savage river that plunged down off our peaks into white rapids and chocolate whirlpools, into subterranean chasms and slimy sunless shadows, into a cañon more vast, more terrifying and more sublime than any other man has known.

To run the Colorado—that was an ambition to send a wiggle up the backbone! Beside it, climbing the Alps and hunting lions in Africa dwindled to tourist frolics. Every year when I was a boy the *Colorado Divide* featured news and photographs of another expedition pushing off. Weeks later we got the report it had passed the junction of the Green and Grand. Months later—if at all—a note straggled back from its few survivors to whet our imaginations through another winter.

Years later, at the other end, we were still electrified by news to keep watch for another party due to be beached along the lower river or spewed out into the mouth of the gulf.

For it had been done. That was the thing! A mere handful of the tens of thousands who swarmed into its basin, a scarce dozen in a century of conquest, had managed to ride the twisting red road of the outlaw Colorado. They were its only travelers; there can be no more.

About the same time, 1825-1826, that Ashley inscribed his name upon the cliffs of the upper Colorado and Pattie was making his solitary way through the cañon wilderness of the middle river, a young British naval officer sailed into its mouth.

The boat was a 25-ton schooner called the *Bruja*, her crew a mixed lot of beachcombers gathered from the tag ends of the world. Their master was young Lieutenant R. W. H. Hardy, R.N., who had entered the Royal Navy in 1806 as second-class boy, and had sailed to Java, Malaysia and Patagonia. Now he had been sent to the Gulf of California by the General Pearl and Coral Fishery Association of London to search for pearl oyster beds. From the Island of Tiburon he had sailed to the mouth of the Colorado in the hope of replenishing his food supply.

On July 20, 1826, the *Bruja* cast anchor in the channel. Next morning when the tide fell she was left stranded on a bar two hundred yards from the river. While waiting for the high tides of the new moon to float her off, Hardy explored upriver. He believed himself, like all those who had preceded him, the first white man to visit the region.

I was now gazing at a vast extent of country visited only by the elements [he wrote]. It is probably in the same state that it was ages ago, and perhaps I am the first person, from creation up to the present time, whose eyes have ever beheld it.

With such responsibility backed by his naval training, Hardy made meticulous soundings and observations, designating on his chart every channel, shoal and island with new English names to replace those of Spanish saints—Sea Reach, Unwin, Greenhithe. At the same time the descriptions of landscape, river and the Cocopahs in his record, *Travels in the Interior of Mexico*—(1825-28), are acute as if made a century later.

He made only one mistake. He assumed that the east fork of the branching he came to upstream was the Gila. What he had found instead were two channels of the Colorado—one the main river, and the other a bayou formed by the high-water overflow from Volcano Lake.

When he sailed away on the early morning tide of August 15th, it was to leave his name presumably on the side channel. For a century it was so regarded: the False, or Hardy's, Colorado. Then in 1909 nature corrected his mistake and proved history wrong in assuming that Hardy had wrongly designated the bayou for the main channel. The Colorado on its great rampage broke through to Volcano Lake and continued on to the gulf by Hardy's channel. Today, still undeflected, it flows over the bar on which the little *Bruja* was stuck.

Twenty-four years later, in 1850, the United States began to explore the river. A military post was established at the mouth of the Gila to protect the rush of forty-niners. Provisioning it by pack train across the desert proving impracticable, a 120-ton transport schooner, the *Invincible*, was ordered to ascend the Colorado for a reconnaissance survey. A prairie schooner would have had better luck. The *Invincible* got stuck at almost the same point as had the little *Bruja*. Luckily her commanding officer, Lieutenant

George H. Derby, had read Hardy's report although the War Department had obviously ignored it. Finally getting his schooner off, he recommended that "a small stern-wheel boat, with a powerful engine and a thick bottom" be used thereafter.

Here now begins the short and fabulous era of Colorado steamboating. It is too humorous to be tragic. In 1851 George A. Johnson, a flatboat freighter, started things off with a little stern-wheeler called the *Yuma*. Next year the *Uncle Sam* was built in San Francisco, shipped in sections and reassembled at the mouth of the river. She promptly hit a snag and disappeared. Johnson tried it again with the *General Jessup,* which blew up under a head of steam—presumably trying to push up a river more silt than water. And now success! The *Colorado,* his new steamer, began to trudge up and down the lower reaches of the dusty red river.

By this time officialdom was furious. Rivers were navigable. They were arteries of commerce, lines of communication—and new military posts needed freighting services far above Yuma.

Q.E.D.

Wherefore the always-practical, efficient and economical War Department got busy. A steel steamboat 54 feet long, with a bow deck to mount a howitzer, and aptly named the *Explorer,* was especially designed and built in Philadelphia. From here it was shipped down the Atlantic, across Panama, up the Pacific to San Francisco, and thence reshipped in pieces to the mouth of the Colorado and reassembled.

While all this was going on, Secretary of War Jefferson Davis invested in another fleet of ships. He secured from Congress an intial appropriation of $30,000—by means of a convincing 238-page report, fully illustrated—for the purchase of thirty-three camels in Egypt and Arabia. In 1856 thirty camels arrived, and in 1857 forty-one more, together with native drivers. Quartered in all the forts from California to Texas, they were put to work sailing over the Great American Desert—a "Camel Express" between Los Angeles and Tucson being one of the chief voyages. It was a several-hundred-thousand dollar dream whose soft padded feet accustomed to the sands of Egypt were quickly torn and lamed by the lava-strewn deserts of Arizona. Finally turned adrift, the camels wandered over the whole basin to survive for many years.

While these "ships of the desert" were stampeding cattle, stray buffalo and wild horses, scaring lonely trappers and prospectors out of their wits, and being hunted down by angry ranchers, that other beautiful dream, the steel *Explorer*, was puffing up a head of steam.

On the high tide of December 30, 1857, she took off upriver. Immediately, in full view of the laughing garrison of the fort, she ran aground. Further to embarrass her commanding officer, Lieutenant Joseph Ives, a report had been received that the Mormons were stirring up the Indians along the Colorado, and the fort commander had just dispatched a detachment of men under Lieutenant White to make a reconnaissance upriver. They went in Johnson's old sternwheeler, the *Colorado*.

Ives stifled his disappointment like a good soldier. To the jeers of the garrison he managed to push and pull the besmirched *Explorer* into water again. When he did take off upriver, it was only to meet the *Colorado* puffing back. Johnson had run her up to Boulder Cañon, covering twice the length of river in his unofficial old tub that Ives's freak steamer was to make. Ives had every reason to be hopping mad, and didn't answer the *Colorado*'s toot as she passed.

He wrote in his report, "Senate Document, 36th Congress, 1st Session": "We were three days accomplishing a distance of nine miles. A boat drawing six inches less water, and without any timbers attached to the bottom, could probably have made the same distance in three hours." Worse still, his crew and distinguished company were visibly affected by the river. The imported Philadelphia engineer grumbled at the Colorado as the "queerest river" he had ever seen. Egloffstein, topographer, had been with that other explorer, Frémont; and Mollhausen, the artist, was "a gentleman belonging to the household of Baron von Humbold, and a member of the exploring party of Prince Paul of Württemberg." Both succumbed to the grotesque and fantastic spell of the river to the extent that their engravings, says Freeman, might have illustrated Dante's *Inferno*.

Nevertheless, Ives worked the boat up to the mouth of the Black Cañon, where it suddenly fetched up against a sunken rock.

For a second the impression was that the cañon had fallen in [wrote Ives]. The concussion was so violent that the men near the bow were thrown overboard; the doctor, Mr. Mollhausen and myself . . . were precipitated head foremost into the bottom of the boat; the fireman, who was pitching a log into the fire, went half-way with it; the boiler was thrown out of place; the steampipe doubled up; the wheel-house torn away; and it was expected that the boat would fill and sink instantly by all but Mr. Carroll who was looking for an explosion from the injured steam-pipes.

Undaunted, Ives and some of the party picked up the pack train that had followed the boat, went on to explore the "big cañon," and reached the Hopi pueblos before returning to Yuma. From there he went on to San Francisco and in 1861 turned in his report. It probably contains some of the best descriptions and finest writing in government files. But war had broken out; he joined the Confederate Army and was killed in action.

Thus ended the *Explorer* and the government conception of the Colorado as a meandering red highway upon which steamboats could kick up dust as they churned through the desert.

But until long after the Civil War, river traffic was maintained between the mouth of the river and the infant settlement of Colorado City, which developed into the town of Yuma. Johnson's Colorado Steam Navigation Company ran a fleet consisting of the *Colorado, Cocopah* and *Mojave*, all stern-wheelers. A new Pacific and Colorado Navigation Company brought from San Francisco the steamers *Esmeralda, Nina Tilden* and *Gila*, together with the barges *Black Crook* and *White Fawn*.

Shipping was long and arduous. Freight and passengers from San Francisco had to be transferred from the sailing vessels to the stern-wheelers at the head of the gulf. From here up the river to Yuma, La Paz and Fort Mojave, it was more an adventure than a commercial undertaking. An Indian was always stationed at the bow to take soundings with a peeled willow pole. Kedging or warping by means of trees ashore was necessary in making bends and crossings. At the times of the bores and high water two weeks were often required for the trip. Possibly no other river service in the world required such light steamers, offered more hazards, and vanished so completely without trace.

Meanwhile at the other end of the Colorado its new legend was being born.

William L. Manly and six other bullwhackers were creaking westward in a train of prairie schooners during the '49 gold rush. Crossing the Continental Divide at South Pass, they happened upon an old ferryboat abandoned on a bar of Green River. It was just the thing in which to float down from the Rockies to the Pacific. In high humor they cleaned out the old scow of sand and shoved off.

At first it was a floating picnic. They made thirty miles a day, "which beat the pace of tired oxen." The meadows were full of elk and antelope. There was nothing to do but shoot and eat, and take turns sleeping while floating down the river-road to California.

Suddenly Manly was awakened; his frightened companions thought the river was suddenly disappearing underground. A vast wall of mountains stood dead ahead. Wrote Manly later: "I told the boys I guess we were elected to go on foot to California after all, for I did not propose to follow the river down any sort of hole into any mountain." Nor did he see "any hole anywhere, nor any place it could go." They had reached Flaming Gorge, the beginning of the cañons and their own troubles.

Within a stone's throw of the cliff, the river turned sharply to the right and swept them along the base of the high red walls. Luckily they ran the first rapids, saw Ashley's inscription of 1825 and added a humorous one of their own. But next morning, before Ashley's falls, they wrecked the ferryboat.

Still joking, they hacked out two dugouts from two pines growing on the cliffs and ran Red Cañon. Next they hit Hell's-Half-Mile and overturned. Even then, swimming for life, Manly's sense of humor did not fail him. Ahead of him a big teamster, unable to swim, was desperately clinging to his overturned canoe. "Walton had very black hair, and as he clung fast to his canoe his black head looked like a crow on the end of a log."

Out of the cañon, they came to a camp of friendly Indians who convinced them by signs of the terrible and impassable cañon below. Reluctantly they accepted pack horses, and on foot trudged across the plateau to fall in with an emigrant wagon train going to California. For Manly it was jumping out of the fat into the fire. The train, lost on a cutoff, stumbled into another fabulous cañon, and Manly was one of the survivors. It is good to think his humor carried him through. Few men, if any, have lived to escape both the Grand Cañon and Death Valley, the two great antitheses of America, and still remember their jokes.

A few years later, in the summer of 1867, three men were prospecting on the San Juan near its junction with the Colorado when they were attacked by a band of Utes. A certain Captain Baker was killed. The other two men reached the Colorado at night, hurriedly built a raft, and shoved off. On the fourth day George Strole was washed off and drowned. Weeks later, on September 8th, the raft with the remaining man, James White, reached the little Mormon settlemen of Callville, Nevada, below the mouth of the Virgin. Three weeks after he recovered, he wrote a letter to his brother describing his experience:

. . . i Went over folls from 10 to 15 feet hie. my raft wold tip over three and fore time a day . . . fore seven days i had noth(ing) to eat to (except) rawhide knife cover. The 8 days I got some musquit beens . . . the 16 days i arrive at Callville Whare i Was tak Care of by James Ferry . . . i see the hardes times that eny man ever did in the World but thank god that I got thrught saft. . . .

Josh ass Tom to ancy that letter i rote him sevel yeas agoe. . . .   James White.

This was an astounding document. If true, it meant that this simple, dazed prospector had ridden a raft through the whole Grand Cañon. It prompted a controversy that spread over the entire country and has filled volumes.

White's stanchest supporter was Dr. C. C. Parry, an eminent engineer and scientist attached to the Union Pacific survey party then working in the region. After talking to White at Hardyville and making an investigation, Parry concluded the raft voyage was true. His report to General William J. Palmer, head of the survey, and to J. D. Perry, president of the U.P., and subsequently published in the St. Louis Academy of Science, remains its best documentary evidence. This gained wide publicity.

Everyone else believed White a liar; not a riverman since but has stanchly attested it an impossible feat.

White meanwhile, with a commendable disregard of notoriety, went ahead making a living. He finally settled down in Trinidad, Colorado, working as an expressman. Simple, honest and straightforward, he never once referred to his great adventure.

Years later, Robert Brewster Stanton, who had ridden down the Colorado and was writing the most comprehensive history of the river, visited him. White then, in 1907, was a hale and hearty man of seventy. Without variation in his original story of 1867, he gave Stanton an identical account.

This Stanton compared with his exhaustive research, which included the sheet of paper on which Dr. Parry had written his original notes when interviewing White. In his famous, unpublished book on the Colorado, Stanton's discussion of the White voyage covers 200 pages. A condensed version of 25 pages was published in the *Trail* of Denver, in 1917, under the title *The Alleged Journey, and the Real Journey of James White on the Colorado River in 1867*. It sums up what is probably the real truth of the voyage:

> Therefore, I conclude from his own testimony supported and proven true, that James White never passed through a single mile of the Cañons of the Colorado river above the Grand Wash Cliffs; but that he did float on a raft or rafts, on that river, in the year 1867, from a point near the Grand Wash to Callville, Nevada, a distance of about sixty miles—where he was stopped and taken off his raft.

That this distance required so long and rendered White "a pitiable object, emaciated and haggard from abstinence, his bare feet literally flayed by constant exposure to drenching water, aggravated by occasional scorchings of a vertical sun; his mental faculties, though still sound, liable to wander and verging close on the brink of insanity" is commentary enough on what a man would have to face in running the whole river.

Nevertheless, his voyage impelled a ranking army officer to recommend to the War Department the building of another large steamboat to explore the river. That same spring, Secretary of War Stanton had received a letter dated March 29, 1867, still more delightful: "Communication from Captain Samuel Adams Relative to the Exploration of the Colorado River and Its Tributaries." This he duly passed on to the 42nd Congress as "House Miscel. Document No. 37."

The letter introduces us to the suave and persuasive Honorable Samuel Adams. It seems that he and a certain Captain Thomas Trueworthy had brought a small sternwheeler to the Colorado and ascended it to the head of Boulder Cañon. Wherefore he hastily wrote his momentous conclusions. "The Colorado must be, emphatically, to the Pacific Coast what the Mississippi is to the Atlantic." He went on to say, "I should have ascended the river farther, but my means were exhausted, the exploration of the last two years and a half being attended with great pecuniary embarrassment to Captain Trueworthy and myself." Nevertheless, he was "satisfied that there are none of those dangerous obstructions which have been represented by those who may have viewed them at a distance, and whose imaginary cañons and rapids below had almost disappeared at the approach of the steamer."

These were sentiments congressmen could understand. They promptly passed the following "Concurrent Resolutions":

Resolved by the House of Representatives, that the thanks of this legislature are due and hereby tendered to Hon. Samuel Adams and Captain Thomas Trueworthy for their untiring energy and indomitable enterprise as displayed by them in opening up the navigation of the Colorado River, the great natural thoroughfare of Arizona and Utah Territories.

Two years later the House received a bill "praying compensation for services rendered," enclosed with a report addressed to the then secretary of war, William Belknap, which opens: "I herewith transmit to you my report respecting the exploring expedition in which I have been recently engaged, the object of this being to descend the Blue River to the Grand, and from thence to the mouth of the Colorado River of the West to the head of the Gulf of California."

What had happened at once removes Adams out of his role of buffoonery and puts him in that more tragic one of plain damn foolishness and extraordinary courage.

He had hiked up to the gold camp of Breckenridge, Colorado, about 11,000 feet high and only about eight miles from the very top of the Colorado Rockies. Here he organized an expedition of eleven men, built four rowboats, and started down this little tributary in the very headwaters of the Colorado.

Descending 4,000 feet in ten days, losing five men, two boats, most of their supplies and being pulled out by a rope himself, Adams reached the upper Colorado (then called the Grand) on July 23rd. Ten days later, still game, he set off again. The prospect was frightening. The walls of the cañon rose over a thousand feet high, the water churning like a millrace, the first fall a sheer drop of "six feet perpendicular." But now the six men were wary, letting down the boats by ropes, and portaging supplies until their boots were completely worn out and several men prostrated. Yet with all their precautions, they lost most of their food the first day.

By August 8th, all the boats had been wrecked. Supplies were reduced to 30 pounds of flour, 6 pounds of coffee, and 12 pounds of bacon. Three of the men left, suffering from excessive exposure.

Adams and his two remaining companions were still game. They built a raft and continued on. When it broke up, they built a second. With the crash of their third, they lost all but a frying pan and a bit of bacon. On August 14th, their fourth and last raft struck a rock and was wrecked. It was the end of their journey down the "Great Natural Thoroughfare of Arizona and Utah."

"Worn out by excessive fatigue and constant exposure in the cold water," Adams wrote simply, "I confess that it was with no ordinary feelings that I was compelled to yield to the force of circumstances."

Dropping a mile in altitude in a distance less than two hundred miles, losing four boats, four rafts, all of his supplies and eight men, he had made a try on a portion of the river that has not been duplicated.

Up to 1869, then, the great red river of the west had been partially run from both ends.

From its high headwaters in the Colorado Rockies, Adams had run down the Blue into the Grand, and Ashley and Manly had descended the Green from Wyoming—the tributary arms of its Y.

The little steamboats of Derby, Ives and Johnson had puffed up the stem from its mouth to Boulder Cañon; and White had covered a short distance above it to the mouth of the Virgin and Pierce's Ferry.

There still remained the entire middle river—the stupendous cañon stretch from the entrance of the San Juan through the "Big Cañon" itself. . . . And even as Adams gave up in the tributary gorges far above, a one-armed professor was sweeping down the length of the main river for the first time.

John Wesley Powell has a unique and enviable record. Enlisting at the age of twenty in the Union Army, he lost his right arm at Shiloh and gained the rank of major. Mustered out of service at the end of the Civil War, he became a professor of geology in Wesleyan University of Illinois. A short time later, while in Colorado and Wyoming, he grew interested in the river. As a result, he took a rowboat ride down the Colorado and wrote a book embracing his findings.

Henceforth he jumps into fame. His detailed report published by the Smithsonian Institution, together with his journal, comprise the classic account of the exploration of the river. Thereafter he organized the United States Geological Survey and was its head for thirteen years, as well as reorganizing the Bureau of American Ethnology and becoming its director until his death. But he is remembered best as the first man through the Grand Cañon.

Funds for the expedition were provided by the state institutions of Illinois and the Chicago Academy of Science. The only aid given him by the government was permission to draw rations from any available army posts—an interesting commentary on that government which had spent hundreds of thousands of dollars building impracticable steel steamboats, lauding wild claims and conjectures, and yet which was to use his own report as its basic study of the river thereafter.

Powell had built to his order four rowboats 21 feet long, divided into three compartments, the two on the ends decked to form watertight cabins. Provisions were assembled for ten men to last ten months, together with scientific instruments for making astronomical, physiographical and meteorological records of the voyage.

Like most of the others preceding it, the expedition elected to descend the longest tributary. The start was made from Green River, Wyoming, on May 24, 1869.

In Lodore Cañon, where the river turns into Colorado, one boat was lost; and at the mouth of the Uintah, Frank Goodman, a young Englishman, left the expedition saying he had "seen danger enough." This was as far as Ashley and Manly had got. From here on they were the only known men to have attempted to run the river.

On August 29th, confronting the Grand Cañon, they abandoned another boat and a considerable number of instruments. Here at Separation Rapids, three more men deserted the expedition: Dunn, a hunter and trapper, and two brothers named Howland. The probable cause of their desertion was the irritable and bullying manner of the major's brother, Captain Walter Powell. He had served in the war, escaped from a Confederate prison, and wandered in the woods for a long time before getting help. This experience had evidently made him a little "queer," and the privations now encountered had accentuated it. The men, however, refused to take any food from the supply stock, and were given a parting salute. A few days later they were killed by Indians on the Shivwits Plateau as they were making their way to the Mormon settlements.

Next day the expedition arrived at the mouth of the Virgin—the first men to run the river through the Grand Cañon. Both Powells turned back to Salt Lake with a Mormon bishop. The four remaining men went on to Fort Mojave: Sumner, Bradley, Hawkins and Hall, to whom is due the chief credit of the expedition.

A second descent was necessary to complete Powell's observations. The great success of the first made it pos-

ITS TRAVELERS 239

sible. It not only brought appropriations from Congress, but supervision by the Smithsonian Institution.

While Powell was making his plans, the army jumped in with another crackpot scheme. Three boats were built in San Francisco, while Powell's were being built in Chicago, and shipped to Camp Mojave. In these Lieutenant Wheeler of the Engineering Corps was sent scrambling upriver through the Grand Cañon. The boats actually made it to the mouth of Diamond Creek. It was an excellent job on Wheeler's part; but why the large expenditure was made, why the expedition was independent from and not co-ordinated with Powell's, what the purpose and results were—and why, in the name of Moses, it was sent *upriver* against rapids that had just been proved difficult to run downriver is still a military secret.

The start of the second Powell expedition of three boats was made like the first from Green River on May 22, 1871, and reached the mouth of the Kanab on September 6, 1872, the last 125 miles being given up because of high water.

Unlike the first, it was a voyage of scientific exploration. For a month or more it would stop to make complete records; the party spent the winter at the mouth of the Paria; pack trains at intervals arrived with supplies. The fact that Powell took none of his former party lends credence to the dissension previously encountered. Included was Powell's brother-in-law, Professor Thompson, a capable geographer and the major's nephew. All were from the East, and all but three had had military service—no chances were taken with independent Westerners.

In the party was a strapping 17-year-old boy, Frederick Dellenbaugh, who was yet to sleep his first night out of doors, being hired as an artist to the geologists. Nearly forty years later he wrote his account of the trip in *A Can-*

yon Voyage which remains the most popular handbook of the river. Every writer, as this one, owes him a debt for a thoroughly complete, accurate and wholly delightful record not surpassed.

What strange vivid little pictures he gives! Major Powell reading to the men from *Lady of the Lake* in beautiful Brown's Hole, from his high armchair strapped to the deck, while floating in still water, or from Emerson "as we slowly advanced upon the enemy." A pack train arriving with supplies and news of the burning of Chicago. The description of John D. Lee, the Mormon instigator of the Mountain Meadows Massacre, who was living at the mouth of the Paria, and of his eighteenth wife, the loved and admirable Sister Emma. Then suddenly the dreaded maelstrom with the apt name of the "Sockdologer"—"one of the most fearful places I ever saw or ever hope to see—a place that might have been the Gate to Hell"—and his accounts of running the rapid. . . . But the book should be read, not quoted.

The Powell expeditions proved the Colorado could be run. Voyagers no longer had the black, haunting dread of being confronted at any turn with a falls high as Niagara, from which there was no escape. The other dangers, more real, still remained, and the wild, intrinsic mystery of the river itself which defied measurement by scientific instruments.

But at least one unfortunate man conceived it as a business venture. Frank M. Brown believed a railroad could be built through the river's gorges from the Colorado's junction with the Green to the Southwest and thence to the Pacific coast. He incorporated this line as the Denver, Colorado Cañon and Pacific Railway, with himself as president, and drove the first stake on March 26, 1889, at Grand Junction, Colorado.

Immediately thereafter he organized an expedition to

make a preliminary survey. There were six boats and sixteen men including Robert Brewster Stanton, an excellent engineer, two guests of Brown's, and two colored servants. Luxury indeed!

Doomed from the start, the party pushed off on May 25th. The first week they lost 1,200 pounds of food. By June 16th all the provisions were gone. Three men left the party, then five more. The rest continued on to Lee's Ferry where fresh provisions were obtained.

On July 9th the remaining eight men in three boats pushed off again. A pall of gloom settled over them all. For one thing, a skeleton had been found; for another, they were entering Marble Cañon where the river seems to be burrowing into the bowels of the earth. That night President Brown "seemed lonely and troubled" and asked Stanton "to sit by his bed and talk." That night Brown dreamed of the rapids for the first time since the start. Next morning it happened. The boat was upset by the boiling rapids. One man was thrown into the current and carried to the bank; Brown was thrown into the whirlpool and disappeared. Only his notebook was seen thrown up after a day's vigil at the spot.

Stanton carried on. During the next three days twenty-four rapids were run without mishap. In the next, two more men were drowned. With only four men left, too few to lug the heavy boats over portages, Stanton gave up the trip.

Returning to Denver, he revived the spirits of the directors and outfitted another expedition of three boats and twelve men. After interminable labor and hardships lasting four months, the party reached the Gulf of California. Still loyal to his chief and his project, Stanton drew up a report etsablishing the engineering feasibility of a railway down the Colorado. But the construction was too costly; the project was given up.

Stanton continued his career as an eminent engineer in Cuba, Sumatra and Canada. But the Colorado stuck in his mind. His spare time, throughout the rest of his life, he devoted to writing a voluminous book covering every phase of the river and its explorers. His 200-page treatment of White's raft voyage has already been mentioned. Records of the earliest Spanish expeditions he laboriously verified to astronomical observations. He even corresponded with the Vatican to connect vague inscriptions on the cliffs with rumors of early Catholic missionaries. The completed work in two volumes, *The Exploration, Navigation and Survey of the Colorado River of the West, From the Standpoint of an Engineer,* was probably the most comprehensive manuscript ever written on the subject. It was not yet published when he died in 1922. Ten years later, however, it was edited down to 232 pages by James M. Chalfant, given a foreword by Julius F. Stone, and published under the title *Colorado River Controversies.*

While government, business and science were adjusting their spectacles for official looks at this outlaw river, a lone Mormon succumbed to its spell. Nathan Galloway was a trader and trapper. For years he had sought out the most remote and inaccessible cañons of the Colorado simply because they retained the most game. Certainly he knew the river as perhaps no other man did.

Few men knew that in 1895 he had boated from Green River to Lee's Ferry. Next year, late in September, he and another trapper, William Richmond, started trapping down to the mouth of the Paria, intending to return overland by way of Kanab. But they were caught by the lure of the river and continued on down to Needles.

Taciturn, solitary and unsung, Galloway said nothing of his ventures and remained unknown. A few years later he happened to be hired as a guide by Julius F. Stone. Stone was the owner of a Columbus, Ohio manufacturing concern, and a wealthy sportsman who had hunted and canoed in Canada. The Mormon trapper was a find. Stone planned an expedition more exciting than a hunting trip to Africa: to run the Colorado and shoot photographs of the whole cañon series. Galloway knew boats and the river; Stone had the money. It was a rare and successful collaboration.

Galloway was brought to Columbus. Four boats were especially designed and built according to his directions—the model thereafter used by all expeditions. S. S. Deubendorf, a Utah photographer, and two other men were selected. On September 12, 1909, the party of five men in four boats pushed off from Green River, Wyoming.

On October 15th they reached the junction of the Green and the Grand in record time. Two weeks later they passed the mouth of the Little Colorado—record time again. And on November 19th, after running the Grand Cañon, the boats grounded safely at Needles. It was the best time ever made, the most rapids run instead of portaged. Not a man or a boat had been lost, and hundreds of photographs had been taken. As a pure sporting venture their feat probably has not been equaled. Especially not by men far past their prime, for in 1911 Deubendorf died, and in 1913 Galloway.

Other trips were made, particularly by George F. Flavell and a companion in 1896; and another in 1907 by Charles Russell, Bert Loper and E. R. Monette in three steel boats covered with wood and canvas. Many others were unsuccessful and anonymous. In the "Graveyard" of Cataract Cañon at least ten men were killed within a few years.

Of all the Colorado's travelers living today, two brothers probably know and love the river best. Both photographers, Ellsworth and Emery Kolb have made it their hobby, business and life. For nine years they lived at the head of

Bright Angel Trail on the rim of Grand Cañon. Their aim was to make a complete photographic record of the cañon. Always in mind was the hope of running the cañon, so in all their photographic explorations they carefully observed the walls and side cañons in case they were wrecked and had to scale the walls to get out.

In 1911 they decided to make it. Before they left they worked out a schedule: Mrs. Emery Kolb and her small daughter were to look down through their telescope in sixty days for their campfire, six miles across the cañon.

On September 8th they left Green River, Wyoming. Seventy days later, on November 16th, wife and daughter saw their campfire down the cañon. On December 19th, after a rest, they continued on to Needles, arriving January 18th, after being on the river 101 days, carrying motion-picture equipment and films through 17 cañons, 365 rapids, and down a descent of 6,000 feet. One of the brothers later went from Needles to the gulf to complete the journey from Green River, Wyoming, to tidewater.

With the 1923 United States Geological Survey expedition, the exploration of the Colorado virtually came to an end. Headed by Claude H. Birdseye, chief topographical engineer, the expedition was organized to make a new map of the Grand Cañon—the last stretch to be surveyed.

There were four boats of the Galloway type. They contained a leader, topographical engineer, hydraulic engineer, geologist, rodman, cook, and a boatman for each boat. Emery Kolb was the photographer. Included in the equipment was a motion-picture machine and a radio. Modern, indeed, to hear on the night of August 2nd of the death of President Harding, and a message to the "engineers braving the rapids of the Colorado!" The start was made at the mouth of the Paria, at Lee's Ferry, named after the Mormon renegade, and where Escalante had camped in 1776 while searching for the "Crossing of the Fathers." On October 20th the party disbanded at Needles.

The outlaw Colorado had been ridden from end to end; it had been surveyed, mapped, photographed, put on paper. It was soon to be curbed. Its legend was dimmed, its physical life laid bare under the microscope. Thereafter the river belonged to the imminent future—a future longer than the river, darker than its cañons, but compounded of its same ancient and enduring mystery.

CHAPTER 7

"The People"

AFTER all these—all the white immigrants from the Conquerors to the Settlers—we are back where we started, back to the Indians.

Now, there are two ways of looking at Indians.

First, as a romantic fantasy. The Noble Savage: Exhibit Number One in America's original three-ring circus, the Reservation. The Vanishing American penned up in his zoo like the dying buffalo—neither a man nor a citizen, merely an incompetent, a dependent ward of a patronizing bureaucracy. But clothed against reality in a sentimental halo. Chief Romantic Redskin in beads and feathers, with war whoop and tomahawk.

Or we can look at him as a psychological fact. A fact bigger than a buffalo, big as life. For the Indian today lies heavy on our national conscience. And he lies at the root of our racial psychosis, which is worse.

Both of these views are sides of the same mirror. "The only good Indian is a dead Indian." For now that we have killed him off we're stuck with his ghost. We're still afraid of him. The white race has funked the Indian and it hates him for it.

Today we have just finished a great and terrible war for survival. A different people, a European race calling itself a "superior" and a "master" race threatened to enslave us. It would have forced upon us a new god, strange customs, a different way of life.

But we are free men. We here in America believe in our freedom—in our own tongue and our right to use it, in our own customs, religion, and traditional way of life. And so we resolved to fight this blond and alien race whose own ancient homeland lies across the great salt water.

The war was not of our making. We were content here to lead our own lives. For a long time we trusted these aggressors who proclaimed time and again they had no territorial designs upon our land. We have observed the solemn treaties made and broken without warning. And we have witnessed the fate of other nations, other tribes, who believed them also.

Fifth-columnists in their midst, those whose tongues speak two ways, prepared the way for invasion. Scattered traders at first, then settlers posing as friends who had voluntarily forsaken their own country for another only to claim violation of their "rights." And suddenly, without warning, with new and frightful weapons, the enemy attacked.

So at last, now that it has happened to us, we know our enemy. An enemy as old as man, limited to no race, no creed, no time. Its real name is Greed and Aggression despite the righteous aliases it uses.

We are talking as Americans. But as Americans of 1944 or 1864, as Americans white or red?

History in the long perspective cannot distinguish which. It is immune to propaganda, to the lie that might makes right. It draws a distinction between us in our roles of aggressor and aggrieved. It does not believe in a just God that can be invoked to jump the fence of principle and back again merely to keep on our side.

Make no mistake about it. Condemn the German Non-aggression Pact with Poland, and we indict not one but hundreds of solemn treaties with Indian nations made by our successive presidents and duly ratified by our Congressses—treaties that were not "Ten-Year Pacts" but for "as long as the sun shall rise, as long as the waters shall run, as long as the grass shall grow."

Condone our advance west from the Alleghenies, through the Six Nations of the Iroquois, through the Five Civilized Tribes—the Choctaws, Chickasaws, Cherokees, Seminoles and Creeks, through the Four Nations alliance of Cheyennes, Arapahoes, Kiowas and Comanches, and we must applaud the Nazi advance into Poland, Czechoslovakia, the Sudetenland and the Low Countries. "Lebensraum!" Or did we want to "free" the Indians from the oppression of their native chiefs?

Bewail Lidice. Then recall the obliteration of a village on Sand Creek, Colorado, in 1864. A village of Cheyennes and Arapahoes, asleep in their tepees in a peaceful land. The leader of the whites who crept upon it was also loose-lipped, boasting and fanatical—the Reverend J. M. Chivington, a minister of the Methodist Church and a presiding elder in Denver. A moment's hesitation. "Kill and scalp all Indians, big and little," ordered this little American Hitler, "since nits make lice." Then it began: what the hardened American General Miles called "perhaps the foulest and most unjustifiable crime in the annals of America."

Without warning, every man, woman and child was killed—75 warriors, 225 old people, women and children stumbling from their lodges. "Among them," records Struthers Burt in his account, "was a three-year-old boy who, left in the deserted village after the massacre, walked out of a tepee and started uncertainly in the direction he thought his father and mother had gone. It took three shots from three different troopers to kill him; and much profanity and laughter."

After the present war there will probably be a board of inquiry to report coldly and dispassionately on the Lidice case.

This is a paragraph from the government commission which reported on the Sand Creek Massacre, and which was headed by the cold-blooded General Sherman who had marched though Georgia:

"Fleeing women, holding up their hands and praying for mercy, were shot down; infants were killed and scalped in derision; men were tortured and mutilated in a way which would put to shame the savages of interior Africa."

Chivington even shot the half-breed whom he had forced to guide him to the village.

What manner of men were these whose crime parallels the worst of the Nazis? Good, average Americans, city dwellers not menaced by peaceful Indians. The most prosperous and respected banker of my home town was one; I well remember his pompous speeches on the subject to us school children. He was always proud of his participation, and because of it was given signal prominence in the town's *Pioneer Reminiscences* so tidily published a few years back. He had helped to "Win the West." As for Chivington, he even ran for the legislature afterward.

At the time of the massacre, the old chief White Antelope was standing in front of his lodge with folded arms, calmly singing his death song. He was murdered by the cavalry who mutilated him horribly.

Reported the commission: "No one will be astonished that a war ensued which cost the government $30,000,000 and carried conflagration and death to the border settlements. During the spring and summer of 1865 no less than

8,000 troops were withdrawn from the effective forces engaged against the Rebellion to meet the Indian war."

But this report was made in 1868, only four years later. The war had just stopped to get its second wind. Ten years more of it lay ahead.

North to their allies marched the outraged southern Cheyennes and Arapahoes. Through Colorado, Nebraska, Wyoming and the Black Hills of South Dakota the drums began to beat. The northern Cheyennes and Arapahoes, the lodges of the Sioux, the Oglalas, the Brules, the Miniconjous and the Crows, the San Arcs and the Hunkpapas—the United red Nations drew together. And their great leaders smoked and talked: Red Cloud, Crazy Horse, Dull Knife, White Antelope, War Bonnet, Two Moons, Red Arm, Black Horse, Sitting Bull, Little Wolf, Black Kettle, Roman Nose.

Thus it began. The red trail of war marked by the Battle of Tongue River, Fort Reno and Fort Phil Kearny, Crazy Woman Battlefield, the Fetterman Massacre, the Wagon Box Fight, and the Rosebud. The trail that began with a massacre on Sand Creek and ended with another on the Little Big Horn. "Custer's Luck"! Gallant, golden-curled, the spoiled and willful Custer and his 208 cavalrymen who were outwitted, outfought and killed to a man.

But the bow would bend no farther. By 1878 it was over. The last defensive federation of free red nations against the white invasion had been defeated. Only in the high hinterland of the Colorado could other separate tribes make a last fight for their homelands.

Deep in Cañon de Chelly the Navajos made their last stand when Kit Carson with Ute guides finally tracked them down for the United States Army. As payment to the Utes, Carson requested that they be allowed to keep the women and children captured.

As a general thing [he wrote], the Utes dispose of the captives to Mexican families where they are fed and taken care of, and thus cease to require any further attention on the part of the Government. Besides this, their being distributed as servants through the territory causes them to lose that collectiveness of interest as a tribe which they will retain if kept together at any one place.

Instead, the last 7,000 Navajos were rounded up for their "Long Walk" into exile and captivity at Bosque Redondo. Released four years later, they made their way back and have increased to 50,000, as Carson warned.

The Utes were repaid by having taken from them over 15,000,000 acres on the Western Slope. Under Chief Colorow they made a last effort to reclaim their land. Defeated, they too were placed in barren reservations in Colorado and Utah.

In Arizona the guerrilla warfare was prolonged by the Chiricahua and Mimbres Apaches under the great leadership of Cochise and Mangus Colorado. Here near Fort Grant took place another Lidice massacre of starving Arivaipa Apaches, most of them women and children under protection of the garrison, by Tucson citizens led by the sheriff. Also another Lezaky, when a band of unarmed Pinal and Coyotero Apaches were murdered under a flag of truce. Placed on a reservation, still another band under Geronimo broke out in 1880 and for six years more ran rampant through Arizona. Finally in 1886 they were captured and imprisoned in Florida.

It was the end of Indian warfare in the United States. Our Lebensraum had been achieved. But the Indians remained. Now they must be "freed" from their land, their tribal organization, customs and religion. Extermination was in order.

Meanwhile several Plains Tribes had been transported to "an everlasting home in Indian Territory." One of them, the Choctaws, gave their name to the new land: "Oklahoma." Another tribe, the Cherokees, had been an independent republic, a recognized nation with a written constitution. Now settled in their new home, they made such progress that Senator Henry L. Dawes of Massachusetts paid them a visit.

He reported:

The head chief told us that there was not a family in that whole nation that had not a home of its own. There was not a pauper in that nation, and the nation did not owe a dollar. . . . Yet the defect of the system was apparent. They have got as far as they can go, because they own their land in common. . . . *There is no selfishness, which is at the bottom of civilization.* Till this people will consent to give up their lands, and divide them among their citizens so that each can own the land he cultivates, they will not make much more progress.

Wherefore in 1887 the "Dawes Act," or "General Allotment Act," was passed. Briefly, it provided that instead of communal, tribal ownership, every Indian was to be allotted a piece of reservation under a fee simple title. They were not expected to increase; therefore the "surplus" was to be purchased by the government for $1.25 an acre and thrown open for settlement.

The rush began. White settlers, land sharks, lawyers and politicians began grabbing up Indian lands for a quart of whisky and a black cigar. From 138,000,000 acres Indian-owned land dwindled to 52,000,000 acres in 1933. Some 86,000,000 acres had been stolen, and 60,000,000 acres more of the "surplus."

The children were being liberated meanwhile. They were forcibly taken from their parents and sent away to white boarding schools. Their hair was cut, they were forbidden to speak their own language, to wear their own clothes and maintain their own customs. They were compelled to undergo religious training by designated Christian sects.

By 1923 there remained scarcely 220,000 Indians in the United States. They had been almost completely exterminated.

"THE PEOPLE" 253

In 1934 there were still only 234,000, including 30,000 in Alaska. Congress then passed the Indian Reorganization Act to civilize and emancipate the Indians. The sale of Indian lands and further allotments to individual Indians were banned, and the tribes were organized into "corporations."

Under this plan the Indians assertedly have made great strides. The Bureau of Indian Affairs claims that there are now more Indians in the United States than there were when Columbus arrived. The number is estimated at 419,000. Congressional investigators deny this, claiming that the figures have been padded to obtain more appropriations for the Indian Office. Actually nobody knows how many Indians there are. There is no official definition of what constitutes an Indian. Census Bureau enumerators in 1940 were directed merely to list as Indians "any person of mixed blood if one quarter or more, or if the person is regarded as an Indian in the community in which he lives."

Despite these asserted "strides," most of the Indians live in poverty. Not more than 2 per cent of all Indian families on reservations average more than $500 income a year. They cannot vote in Idaho, Washington, New Mexico and Arizona. They cannot own, sell, mortgage or lease land on the reservation without consent of the Indian Office. What are their prospects for the future—to continue to be kept racially segregated as museum pieces, or are their inadequate reservations to be abolished and the people finally absorbed? No one knows. There is not yet a clearly defined, long-range Indian policy.

Under such conditions the Indians themselves have been of little help. Stubbornly they cling to their own old ways. School graduates drift back "to the blanket." More and more their racial characteristics of stubbornness and secretiveness have been intensified. Nothing of their ceremonialism, their legends, their profound and subtle religious tenets—that rich indigenous culture of America we need to absorb as much as the Indians need to adopt our own—will they divulge to whites.

So that today the Indians comprise scarcely 2 per cent of the national population, and this minority is virtually imprisoned on its wilderness reservations. Politically, economically, socially, they seem inessential and unimportant. But psychologically the Indian is today still a tremendous fact on this continent.

Why? What has his defeat done to the white?

From the very start, as Carey McWilliams has pointed out, the white had no idea of amalgamating with the red. Always it was a war of extermination. Year after year, mile after mile westward, this idea became more fixed. Indian men, like deer, were called "bucks." From this they became "pesky critters," game to be shot, vermin to be exterminated. In Colorado, legislation was offered placing bounties on the "destruction of Indians and skunks." In Oregon a bounty was placed on Indians like coyotes. They were trailed with hounds, their springs were poisoned. Women were clubbed to death and children had their brains knocked out against trees to save the expense of lead and powder. The clergy sanctioned such practices against these dangerous "seeds of increase." In California, extermination of Indians was approved; American miners, unlike the earlier Spaniards, needed elaborate machinery and the Indians were of no use as slaves. Even Francis Walker, United States commissioner of Indian affairs, stated that he would prefer to see the Indians exterminated rather than an amalgamation of the two races.

So, tragically, from the first meeting of the two races, the white was obsessed with the need of killing off the red. The obsession grew. It became a fixation.

Why? Fear was the great factor. First, the physical fear

of the wilderness, of the great unknown—of which the Indians were a part. Then, later, the psychical fear of the invisible forces of the immense hinterland, of space and altitude, its power-of-place; the psychical fear first experienced by the solitary trappers after they had conquered the physical landscape. For the Indians, understanding the psychical as they did the physical—the spirit mountain which was an inevitable counterpart to the material mountain—were a part of these forces too.

We were deathly afraid of the great mystery of the new, unknown continent. So we conquered the land and overlaid it with the greatest material civilization known to man, ignoring its secret and invisible forces. And we conquered and exterminated the people of its soil, refusing to amalgamate with them in the physical flesh and ignoring their psychical counterpart—their essential spirituality.

The result has been the formation of our white mores violently antipathetic to mixture with any dark race. It began with the Indian, carried through to the Negro and the Mexican, and held with the Asiatic. America as a melting pot holds true today only for white European races, not dark.

It is a tragic psychosis peculiar to the United States alone. The whites in Mexico, the Spaniards especially, amalgamated with the Indians to form a new race, the mestizo. The French allied with them to form another mixed breed. In Brazil, predominant in the world for racial tolerance, the Portuguese and Italians have mixed with both Negro and Indian. Chile is almost totally mestizo. And this holds for most of Central and South America. But in us here persists this national psychosis; the unconscious fear and hatred of the Indian.

Elsewhere he is becoming a physical problem of continental proportions. In all of Middle and South America only Argentina, like the United States, has decimated its indigenous people. Of Mexico's 19,000,000 population, 15,000,000 belong to Indian Mexico—"a nation within a nation." Peru, Bolivia, Ecuador, Colombia, Venezuela and Guatemala are preponderantly Indian. Roughly, over 30,000,000, or one-fourth of the entire population, below the Rio Grande is pure Indian. More than a third of these use their own tongue exclusively. From deep underground this socially and economically submerged bottom layer is steadily rising with a rapidly increasing birth rate. Not even the Catholic Church has effaced what is essentially Indian. For this mass the Spanish-Catholic culture is but a thin patina.

The top layer—the pure white—comprises scarcely one-fifth of the population. It is steadily vanishing; the birth rate is falling. The remainder in-between is becoming ever darker. Nearly 80,000,000, two-thirds of the entire population of Middle and South America, is darkened with Indian blood.

Haya de la Torre, leader of the great underground revolutionary movement in Peru, "affirms the necessity for creating a nation on the basis of Indian elements." His concept of such an Indo-American nation or bloc of nations reinvokes Bolivar's dream of a united Peru, Ecuador and Bolivia, says Beate Salz. This "Indian problem" is the basic problem of the future. To study it a mixed commission has been proposed in Ecuador and Colombia. In Bolivia the first Indian "Congress" since the fall of the Inca Empire in 1533 has just been held. Of the 1,500 delegates only 300 spoke Spanish; the rest, all heads of ancient Indian communes, spoke Aymara and Quechua.

Mexico has set a precedent; her whole agrarian policy, the result of a 30-year revolution, is aimed at giving Mexico back to the Indians. José Vasconcelos predicts the coming of a fifth race, an American race, composed of Indian, Negro, Spanish and Nordic elements.

Even in the United States the negligible Indian population shows the continental trend. It is the fastest growing racial group in the country. For the next half century the expected rate of increase for whites is 19 per cent; for Negroes, 50 per cent; for Indians and Mexicans, themselves part Indian, 139 per cent.

Yet in all the three Americas the United States alone is a white island in a dark and rising sea. Euro-America with its rich materialistic civilization still building upon the old concepts of crumbling Europe; with its racial hatreds, its national psychosis, its deathly inhibited fear of the Indian it almost destroyed.

Here at last the rolling wheel of its pioneer-emigrant has turned full circle.

Only against this vast background do the Indians of the Colorado River basin show in true perspective. In all our America they are the only tribes who still remain rooted to their ancient homelands, whose city-states remain unchanged. Here alone you can glimpse the truth that has given the Indian his miraculous tenacity, his indomitable stubbornness to persist unbroken through defeat.

Go to a remote trading post anywhere in the heart of this vast river basin. The trader will not be surprised to see you. He'll know that if your car isn't stuck to the hub in sand a few miles off it's bogged down in adobe at the last wash. The last arrivals are already in a lather about hiring horses to pull their own out, while the womenfolks are in a dither to buy that perfectly stunning bracelet there in the case. (Sorry, lady, she's pawn!) In one corner swilling warm soda pop stands a bunch of visiting Apaches, brawny fellows with pigtails hanging down from under their big black Stetsons. In the other, cooing softly as doves, loaf two docile broad-cheeked Hopis. Around the walls patiently squats a rainbow of Navajo squaws in cerise, turquoise, yellow and sage-green velveteen blouses, picking lice from each other's hair.

... When in walks the kingpin of creation, with a first mortgage on America itself. You know it the instant he steps lithely in the door. Ragged, sweat stained, a dirty colored headband holding back his uncut hair, he carries about him an incontestable air of arrogant superiority that makes you feel a foreign intruder. You!—a successful businessman, a substantial property holder and an honored citizen of the World's Greatest Democracy! It makes you want to kick this penniless, landless and ignorant savage in the seat of his ragged denims.

But you don't. It isn't because—well, really, fellows, they're tricky, don't know the meaning of fair play, and besides your wife was there. It's simply because you feel like dirt under this Navajo's feet. He doesn't even deign to notice you. And in his own sweet time, without even lighting a Murad, he nonchalantly turns back out the door. Leg over his waiting sorrel, riding on the point, he slips swiftly away with the wind.

Damn! There went something! You can't quite figure what. But long after you've forgotten the truth you sensed in him you'll be willing to bet a Navajo's lordly arrogance against a Mayflower lobbyful of monocled foreign diplomats any day in the week.

The Navajos' name for themselves is "Dinneh"—"The People." THE People, please. Something like a cross between a Tidewater Virginian, an old Charlestonian and a Kentucky bourbon colonel; or an Astor and a Vanderbilt among the Financial Four Hundred. Only not decadent, movie style, like them. This is the real thing.

The Apaches call themselves "Inde" or "Tinde," meaning "The People."

Similarly, "Dzi-tsii-tsa"—"The People"—is the real name of the Cheyennes. "Cheyenne" is but the corruption of the Sioux name for them, "Sha-hi-yena"—"The People of a Different Speech." Even the Arapaho name for themselves, "Inuniana," is simply "The People." "Oklahoma" itself is but a Choctaw expression meaning "Red People" or simply "The People."

There is the land and there is the people. It is as simple as that. And as complex. For the land is the ultimate source of all life. And the physical manifestations of all this life—that of the living stone, of bird, beast and man—derive from it in a continuous evolutionary chain, each link embodying all the principles of the forms that preceded it. "The People," then, is synonymous with that other ceremonial name for themselves, "Made-From-Everything." For they were created last, out of the whole substance of the universe and embodying its corresponding spiritual principles. By this simple term, The People, they both acknowledge the earth from which they originally sprang and their pride in being the consummation of its expression.

In this sense they possess the only true aristocracy which also ultimately derives from the land. America is their spiritual as well as ancestral heritage. No other race residing on its soil has yet achieved their harmonious relationship with it. Indeed, "hozhoni," the Navajo word for happiness, means simply full harmony between man and land. And only in the Colorado River basin has this been perpetually maintained, their one great truth that we have yet to learn.

In the middle of it, in the Four Corners district, lies the 25,000 square mile sand and sagebrush desert of the Navajo reservation.

In the middle of this rise the three mesas holding aloft the Hopi pueblos immemorially protected by their sheer high cliffs, landmarks now as in Escalante's time: Walpi and Sichomovi; Mishnonghovi, Shipaulovi and Shumopavi; Oraibi, Hotevilla and Moenkopi.

To the east are the Jicarilla Apaches of northwestern New Mexico—both those known locally as the Llaneros and the Olleros; and the Mescalero Apaches of southern New

Mexico. Farther east are the fourteen pueblos of the upper Rio Grande.

To the north and northwest are White River and Uncompahgre Utes, and the Paiutes of southern Utah.

To the south is Zuñi, ancient Cibola, huddled close to sacred Corn Mountain, and the largest pueblo today. And Acoma, the Sky City on its high cliff, contesting with Oraibi as the oldest continuously occupied town in the United States. Farther south are the Tonto, San Carlos, White Mountain and Chiricahua Apaches of southeastern Arizona; and the Pimas and Papagos who flow over the border to mix with the Yaquis of northern Sonora.

And to the west, the Havasupais, Hualpais, Suppais and Wallapais; and the once huge club warriors, the desert

Mojaves, Yumas and Cocopahs, few and decadent as they are.

To any boy grown up anywhere in this high cupped basin of Indian earth, Indians are as essentially a part of his life as anything he will ever remember. They may be romantically idealized, as are locomotive engineers and city gangsters—and to far more reason when he sees them in the great myth-dramas of their ceremonials. They may be psychological facts of prime importance in his adolescence, especially if his contact with them is close. But above all they are people.

Myself, I can't remember when I first knew Indians. A half-moon of their dark ruddy faces forms one of my earliest horizons. The faint steady beat of a distant drum is what I hear when I put hands over my ears, as a mariner's son hears the sound of the sea in every shell. To us a worn Navajo blanket on the floor and a cracked piece of Hopi pottery were as inevitable as the hooked rag rug and conch shell of old New England families.

But I remember my first Plains Indians. That long loping crescent of Cheyennes! Naked to the waist on their spotted ponies, with their beautiful headdresses of eagle feathers tipped with red yarn, their war paint and their yells! They loomed suddenly out of the vast empty plain as out of my deepest memory; stood etched a moment against the rampart of mountains as a vision of something precious and worthless and forever enduring; and then vanished instantly from sight—mine and America's too.

They were probably riding by to hire out to Buffalo Bill for the afternoon's performance—though, peculiarly enough, the Wild West Show itself left no trace on my mind at all if indeed I saw it. For I recall them that night when, paint and feathers off, they were reduced to their familiar status of mere men. They sat around the fire carving meat from the spit. Our vegetable huckster, a Cheyenne, was one of them. Next to him sat my father in the circle, himself in turn deftly laying back the flap of fat with the long naked blade he always carried. On small-boned feet encased in meticulously shined shoes he was squatting cross-legged as they; his sinewy brown hands sensitive as theirs in flamelight; his dark face with its great Roman nose jutting out from his celluloid collar.

Till I was ten the mountain Utes came down the pass every summer and pitched their tepees west of town. Buckskin Charlie was chief—a fine-built man, solid and majestic. A little dumb, people thought, to pun his taciturnity. Later we found out one of the things it hid. Old Chief Ouray— for whom a mountain, a town and the son of my friend Ralph are named—had died years before. Buckskin Charlie, his successor, had hidden his body in a cave and steadfastly refused to reveal its location to authorities. Nor did he until long after I had grown to manhood.

His wife Chipeta we then thought smarter. A town is named for her too. She deserves it if only for this example of her shrewdness.

The Utes were in such straits that even their grafting government agents grew worried. An agent then, on $1,500 a year salary, was normally expected to retire with a fortune after three or four years. But Chipeta persuaded them to advance the tribe enough money to fatten their sheep to sell. Several months later the stock cars drew up on the nearest siding to receive the promised sheep. Chipeta directed the loading of the cars. A few days later they arrived at the Denver Union Stockyards, and the seals were duly broken in presence of the government agents. Out came a wobbling parade of the scrawniest, half-starved, bony goats that ever went to a tallow factory. Chipeta and her Utes meanwhile had fled farther back into the mountains with all their fattened sheep to build up a larger flock.

So suddenly it seemed the Utes stopped coming down the pass; no longer each summer their smoke-gray lodges loomed against the cottonwoods. Instead, regularly as they, came the Osages and Cherokees. How different this!

Oil had been discovered in Oklahoma on the stomping grounds of the Five Civilized Tribes. So every summer by the trainful they poured into town, uproariously, sadly, fabulously rich. I used to meet them at the train and herd them to the most fashionable tourist hotel—blankets, rolled-up mattresses, stalks of bananas, pop bottles and all, filling up a fleet of hired Packards. Money to the Osages always assumed its only real significance. It was merely green paper. They carried it in rolls big enough to choke a cow. For a ripe apple one of them would peel off a hundred-dollar bill, expecting no change. Or buying a whole bolt of Chinese red silk, they would start a row because the proprietor angrily demanded more than a one-dollar bill. Unluckily these latter were few; fresh-minted Colorado-silver wagon wheels were the style. But money finally killed them. Not the lack of it. Too much of it.

The old story is true of a bunch of rich Osages freshly registered in the bridal suite in the morning, and timidly informing the clerk late that afternoon that their "spring" had gone dry. They had drunk all the water out of the toilet bowl. Ed Tinker tells it himself. He was one of them. There is still a better one on him. Years later I went on a trip with him into southwestern Colorado. Ed was hunting a piece of land long sold to the family sight unseen by white land agents. On it he figured to build a cabin and keep a cow or two to stave off that national wolf, the depression. We finally found it just south of Durango's "moving mountain" but couldn't get to it. It was the flat summit of a bare, volcanic mesa whose unscalable walls rose sheer from the plain.

Oklahoma will long remember the Osage Tinkers. It was Ed's brother, Major General Clarence Tinker, who was appointed commander of the Pacific Air Force after Pearl Harbor and went down at Midway. He, like thousands of others, is an example of that untapped wealth America ignored in its greed for oil, but which remains our greatest heritage.

Yet the passing of the Osages as a tribe constituted a shameful, tragic period I experienced with embarrassment. They had been forced to trade their birthright for headrights; and cut off from their land, without nourishment, they swiftly withered.

Their antithesis are the major tribes of the Colorado River basin, the Navajos and the various Pueblo groups. Indomitably rooted to their land, they remain integrated as ever. And them I have known best. Through twenty years and more they have given me not only my best friends, but America itself.

Fortunate indeed is he who has received such a rich heritage. For he knows that their land is also his. His body, like theirs, is built out of its native corn. These great peaks rent with storms and arched by rainbows can never quite be explained away mechanically. Old Man Coyote, Spider Woman, the Grizzly-Bear-Who-Married-the-Girl-Created-From-an-Aspen are not strange to him. For his fairy world is not that of Europe. He owes no allegiance to King Arthur, nor to Zeus on a Grecian peak. The voice of mighty Manitou is the voice that fills the silence of these cañons, speaking from these ancient mountains that have always spoken to a listening people. It is the voice of America itself, with its own rhythm, its own deep wisdom, and its own symbolism for the universal truths of living man.

Once I took a horseback ride through another Indian country. It took three months. I rode south along the western slope of the Sierra Madres in Mexico. Traveling from one remote village to the next with an old man or a boy for a guide—the only ones who could be spared from the fields —I passed through the Yaquis of Sonora, the Mayos of Sinaloa, skirted the Huicholes east of Tepic, and into the

Tarascan lake country of Michoacan. Only as the successive idioms changed, when tortillas became chusquatas, was I really aware of the change of tribes. They were all one people. Indian earth and Indian corn, they are the same wherever you find them.

Still later I was reassured of their persistent, subterranean solidarity. This time I drove through Mexico with Tony from Taos. In his lap as usual was his small drum, and down his broad back hung his two long pigtails wrapped in their brightest ribbons. With us also was his good medicine.

It got us forded across slimy soap-green rivers, pushed up steep rocky roads to villages high in the Sierras, opened to us remote little plazas hewn out of the jungle. In the dim spluttering flare of ocote torches at night surly serape-clad men who habitually avoided whites as a plague and professed to know no Spanish, crowded us for hours with interminable Indian talk. In the bright hot noons shy dark strangers brought little presents of seed corn and fresh dried leaves of tobacco. We wanted gourds, and with friendly smiles they offered guajes with handles to fit the hand while dancing.

Then Tony mentioned casually that he would like to buy a few parrot feathers. The news spread; next day they started coming. At any hour we would be aroused to meet a strange Indian waiting with a feather. Not any one would do. Each was carefully examined for quill, color, luster. The interminable dickering went on and on, punctuated by sign language written with expressive hands. There they sat. Tony in fresh shined boots, his long hairbraids falling down over a clean white shirt. And squatting beside him a ragged campesino with dirty toes sticking out of his worn huaraches, an orchid stuck behind his dirtier ear. Not a meeting of Americano and Mexican, of rico and pobre, but of Indians. For centuries, long before the Spaniards came, this identical trading had gone on. The parrot feathers worn in all the great sacred dances up north have come from down here in the jungles of Mexico. They are a symbol of the subterranean solidarity of all Indian America, still mute but powerfully persisting under a white veneer.

And so at last we pause before the still unshaped future. That future lies neither solely in our hands nor in theirs to shape. It is not wholly a social problem. It embraces not alone the relationship of man to man, but of man to nature. And it is precisely this latter which has differentiated the white from the Indian. Nowhere has this difference been more accentuated than here.

As Hugo Fernandez Artucio has pointed out regarding the Incas of Peru, these vast towering ranges have always constituted to the Indian the immediate and the familiar, unseen with the perspective of time and distance. They have been lived with in the direct relation of the particular to the universal, of the perishable to the eternal.

Hence the Indian as an individual never existed. Dwarfed by the immensities about him, he existed as a member of a tribe, a pueblo, of the universe itself. And he expressed himself in forms that corresponded to his collective personality. Property was owned communally. Religious rites were participated in communally, taking the ceremonial forms of great myth-dramas, huge tribal dances and sings. As the pueblo itself rose tier on tier in the semblance of the peak behind it, so was the sacred kiva a miniature of the world seen from within; its round walls the circular horizon, its dome the sky above, the sipapu the place of emergence from the previous world below. Likewise was the Navajo dwelling, the hogan, an icon in its truest meaning. With its door opening to the east, its octagon walls laid with the four directions and the points between, it too constantly oriented him to the greater universe within which he also dwelt.

Indian life, then, like the mountains, expressed itself as

an undifferentiated human mass whose evolution corresponded with the same laws of nature.

One day a strange man arrived—the white man. Individualistic as the Indian was communal, he confronted a psychological nightmare. There were these wild overpowering mountain masses terrifying in their vast anonymity. And there were these people just as incomprehensible and frightening with their vast impersonality.

We know what happened. The white went beserk in his fright. Maddened by his need to express himself, with a will to dominate, he fell prey to his own greed and cruelty. From his onslaught the life of both land and people subtly withdrew. And the Indian, almost exterminated, was left as he is today—hopelessly secretive and withdrawn into himself, the ghost of a collective personality hemmed in without means of full expression.

So today, in its broadest terms, we have reached the apex of our struggle. Both have reached their last oubliette—the white with his monstrous individualism and the Indian with his vast impersonality.

Thus history here as everywhere has been simply "the record of a species of migratory animals called Homo Sapiens—of their impact upon each other and upon their natural surroundings, and of their search for the meaning of the universe."

Over four hundred years ago, long before Cortez planted his first cross of conquest on the shores of the New World, strange signs and drums, prophecies and earthquakes heralded his coming. He came, and the whole continent was dedicated to a new destiny.

Today the drums are beating again. A new and belching volcano has suddenly and miraculously risen out of a sleeping cornfield. A new god is being born. A new race is rising out of America with a culture and a civilization the world has never seen. A god, race and civilization neither white nor red, neither blindly materialistic nor spiritually withdrawn into itself.

This is our new destiny.

CHAPTER 8

The Inheritors

A<small>LL THESE</small> have made us what we are. The land we have inherited. And the human heritage of all those who have preceded us.

Human history is but the story of man's adaptation to his environment. The deepening relationship of a people to their earth. That is the essential truth we read here in this vast heart of America, the upland basin of the Colorado. In the conquerors, whom it defeated. In the padres, whom it rejected. In the trappers, who overcame the land physically and were caught by it psychically. The secret of its hold upon us is the treasure the prospectors sought. Until we find it—the profound and haunting secret of the reciprocal relationship that must exist between man and land—we will still remain outcasts. Outcasts like the Mormons from the rest of mankind in their time, and outcasts from the settlers in turn like the individual outlaws.

For to us, as to them, the haunting cry of space still and forever rings in our ears. The same vast loneliness engulfs us. We have spanned the horizons and the seas three thousand miles apart. Our towns dot the valleys, bestride the rivers, cling even to the mountainsides. Still tormented, we are driven forward and back by a gnawing restlessness.

We can find no peace, no rest. The immense space conquered outside has only crawled inside us. The corrosive loneliness still eats within us. The shriek of a train whistle across the midnight cañon, like the shriek of a cougar—that is the voice we know best, and it is the voice of all wilderness America with its still unconquered space and our secret unassuaged loneliness.

Still the tension keeps mounting within us. Like the Mormons we are driven to a fanatical puritanism. Like the Penitentes to masochistic self-flagellation. And like both to sacrificial murder and crucifixion.

We cannot abide ourselves. How, then, have we lived with each other? With the development of our peculiar racial psychosis we killed off the red, enslaved the black, erected barriers against the yellow, discriminated against the brown, and now find it difficult to tolerate the Jewish race. Only to the white has Liberty held up her torch.

But instinctively we have felt the land to be at the root of all our trouble. A strange New World not to be wooed, won and finally understood. But a new wild land to be quickly overpowered, raped and gutted. Thus we have leapt to the one task by which we could expend the full force of our power, and find a momentary release from our constant nervous tension. To conquer the land and expend its riches.

It is done! We have sunk in it the foundations of the greatest materialistic-mechanistic civilization ever known. On it we have raised the highest triumphant towers ever built by man.

And now we but recoil upon ourselves. For of all men we are the only true killers. The most tolerant and good-natured, the slowest to be aroused, we are passionless, dispirited killers. Of what? Quién sabe? We only know that there lurks in all of us a cold murderous instinct that unannounced and unpremeditated lunges suddenly out of the easygoing shell of our lives. It strikes those we love as often as those we hate. Why not? It is undirected, impassionate, impersonal. The quality of a true half-breed, to strike *out*, not *at*.

And so we topple over our tall shining towers as soon as they are built. We claw down and stamp upon our idols the moment we have raised them to their brief acclaim. Nothing is safe, permanent and secure from our eventual knife thrust. It is as if at last we find even our success no more than a barrier across the path of our invisible destiny. Like the trappers, we are eternally driven by a strange compulsion from the known and loved to the feared unknown.

And deep in our hearts we know it. For all our vaunted practicality, we are the most mystical people on earth. From our souls we see, the pain is ever with us, and so we shout it down with a mammoth jest—the immense and ironic humor of America, to which nothing is long sacred.

Yet all the time the counterforces of salvation have been insidiously working in us against our will and knowledge. The strange and subtle forces of the land itself. Every anthropologist of merit, including the oft-quoted Ales Hrdlicka, has been pointing out for years how "climactic and other influences are gradually working to produce in us some of the facial characteristics they had already produced in the Indians." After a few generations here even the skull types of various pure Europeans change to that of the indigenous Indian. Not only physically but psychically a change is taking place. Jung has stated that when we reach the bottom of the American psyche, we find the Indian.

Slowly but ineluctably the land is rooting us to its soil. First the solitary trappers as individuals, then the isolated settlers as groups, and now us as a race. This is the real history of the vast heart of America, the upland basin of the Colorado. The story of man's adaptation, psychically as well as physically, to his environment. The deepening relationship of a people to their earth.

For a mere handful of men had confirmed us to a new destiny. They had wiped out the past. They had set us on a new trail, rooted deep in the immortal, impassionless American earth. But neither they nor those who followed them had gone far enough. They had not fully achieved that harmonic, reciprocal relationship with the land attained by the indigenous peoples whose trail to it they found.

The end of it is what we seek. This is what we know deep in our loneliness: that we are still strangers to our earth. Like the trappers we know it.

But they have set us on the trail—the long, long trail where the pony tracks go only one way. Nothing for long can hold us back. Neither our own materialistic-mechanistic success, our lofty towers, our own loves. Impassionate killers, we must at last destroy even the incompleteness that holds us back.

And so we look into the mirror to see the meek who have inherited the earth. We need such "a feeling of continuity in our experience as a people, a sense of the past as a living reality conditioning the present." But America on the whole has lived too fast to think.

The industrial East, quickly populated and still linked umbilically to Europe, has lost all perspective of time and place. Everything west of Chicago has seemed remote, irrelevant and provincial. At best it appears strangely picturesque —a romantic escape for a summer vacation.

Always the booming West Coast has been the end of a journey across empty, unregarded space. It still looks westward, not backward, with longing to continue the journey across more empty space.

The era of the Pacific is beginning, just as the era of Europe is ending.

But between them lies the crossroads of the world, America. And now, if ever, America must emerge: its full blossom, its final pattern, and its ideal of a completely synthesized people one at last with its own physical background. In this final synthesis the vast hinterland—the still empty, backward, wilderness basin of the Colorado—must play its important part.

But let's face it. Cowboy boots and high-peaked Stetsons, as well as war bonnets, are out of style. If there is anything that gives me a pain in the neck it's the professional Westerner. You know the type. The senator who insists on being photographed in a ten-gallon hat and astraddle the hood of his Packard limousine when he arrives in Washington. The strong, silent man of the open spaces always on the guest list for cocktails to parade his sunburn profile in front of visiting tourists. The artist who goes native in Mexican huaraches and a small adobe littered with expensive serapes but without a toilet. Particularly the writer who makes a cult of his sectional background. They are all phonies; you can spot them a mile away.

Concurrent with the recent world-wide war there has come inevitably a recurrent epidemic of nationalism. America, for its part, has been shaken to the roots. All the self-banished aesthetes have come scurrying home from the Riviera. Desperately we are hanging on against the high tide of change to what we are and to what, by Jesus, we intend to remain. Hence we are ready victims of the epidemic's most virulent attacks of rampant sectionalism.

Movies, radio and book publishers, theaters and slick-paper magazines are all afflicted. Anything American goes. Especially if it goes back far enough. What is it we want to see? Here in the Wide Open Spaces of God's Country we want to see ourselves as the rootin'-tootin'est, hell-roarin', God-fearin' sons-a-bitches that ever were suckled on panthers' milk and cactus juice or raked spurs in a hoss. Oh, but we have our tender side too! We sure do, ma'am, beggin' pardon. Why, when the old moon comes up over Pleasant Valley yonder, and the night breeze comes astealin' along the ridge like a pesky coyote, it makes a fellow think how good it is to be back here leadin' a noble simple life like his pappy and his ma afore him. A real American. Nothin' fancy. Just plain square-shooters, us folks.

Listen to the senator. Look at the movies. Hear the radio. Read almost any magazine, almost any book on the West. The same old stories. The same fictional vernacular. Pure, unadulterated hokum.

Why?

We insist on seeing ourselves as we like to think we were. We're afraid to see ourselves as we are.

Is it because we have lost our innate sense of direction? Or are we so afraid of the future that we cling to a false romantic past?

Let's face it. Cowboy boots, peaked Stetsons, war bonnets and huaraches are out of style. The professional Westerner is barking up the wrong tree. He's making a fool of himself and us too.

Dress, manner, vernacular and all the other affectations are all phony. Not the psychology of the Indian, Penitente, Mormon and all the others reflected in us. Let's keep this straight.

But even this latter is changing. The New Deal, Frank Sinatra, the Depression, the WPA, Clark Gable, the War, Coca-Cola, refrigerated beef, the Petty Girl and Donald Duck have left their ineradicable marks on our very souls. Not long ago I took some friends far up a remote mountain cañon to attend a Saturday night dance in a small Spanish-American village. I wanted them to see the old, graceful varsoviana. It was not danced. The people were too busy dancing to the boogie-woogie.

I was in California at the time of Pearl Harbor. A sudden telephone call sent me rushing to an Army Induction

Center. An Indian boy who had been drafted had attacked a group of majors and colonels because they insisted on cutting off his hair. Imagine that indignity! Thousands of Indian boys have been bereft in an instant, by a clip of shears, of generations of tradition.

What about the old folks at home?

From one pueblo a group of old men came to call on a friend of mine. They wanted a letter written to President Roosevelt. "It is spring," they said. "It is time to plant our corn. But there is no one left to plant it. All our young men have gone to war. They have been gone a year. Now it is spring and time to plant our corn again. But they are not here to plant it. Please send us back our young men. They have been fighting long enough. It is the corn that must be planted, you understand. It is not us, their wives, their mothers, their own small sons. It is the corn, this you will say."

A year later they still sat, the corn unplanted, munching on the dollars the government sent them from their army sons in Guadalcanal, Tarawa, Burma, Iwo Jima, Normandy, Germany. When all these sons return, will it be back to hairbraids and the blanket, to finally plant the little milpas of squaw corn among the pines? Or will they have learned the possible persuasiveness of the vote, the almighty power of the dollar, a new trade, a larger brighter vision?

So it goes among Spanish-Americans and Anglos too.

Whoopee! *Beat Me Daddy Eight to the Bar* from a juke box instead of *La Paloma* on a guitar. Mother wants a high-altitude pressure cooker to can her own vegetables for winter. Father has bought another Ford; he's going to try this contour plowing. But brother is nuts about airplanes. Sister does her hair like the movie actress of the month. Baby doesn't believe in Santa Claus. Watch de cops when dey move in on de gang. Rat-a-tat-tat-tat-tat! Who the hell was Billy the Kid anyway? Give me Superman. *Rum and*

Coca-CO-la! What! Me work my tail off for that damned little Schicklegruber for ten bucks a week like I been makin'? Who, ME? Not when I can grab off seventy-five over the hill. OH-oh! Where's the fire sale, honey? Nice pair of gams if they was gettin' you anywheres.

What's cookin' anyway?

It's about time the professional Westerner and all the rest of us were finding out.

Pyramid City

What is cooking, to be precise, is that the senator's half million constituents in 150 little towns and villages are building up the quarter million square miles of Pyramid City for occupation by, say, ten million future residents.

Don't get them wrong. There's nothing romantic about it, any more than there was about founding it some time back. The Stetson-hatted senator's still hell-bent on getting his proper cut. The huarache-clad artist is going to paint over the sweat and pain with a glossy finish of glamour. And inevitably the romantic-western writer is going to glorify the wrong heroes. But the people are going ahead with it just the same, taxes or not.

A great city is the most fascinating thing alive. Thomas Burke, who never set foot outside London, wrote twenty books on his travels through its streets. He did not begin to cover the surface of the subject. Since time immemorial youth has followed its dreams and ambitions to the turrets of Camelot, the garrets of Paris, the towers of New York. It is an idea rooted in every heart, a shadow cast upon the most remote peasant. The tremendous idea, the immense shadow of the city.

And so likewise has the city become the most corrupt, the most cancerous sore upon the body of civilization. All the strength of far provinces has been required to maintain it. Whole populations have been enslaved to stoke its furnaces. The land itself has been sucked dry. It is not enough! Inhuman and insatiable, a monster with a hide of concrete and guts of steel, it dominates humanity itself. It creates the slumdom of the poor and the boredom of the rich. It distorts all men equally, both those powerfully isolated in their tall towers and those groveling in the sunless gloom of their deep cañons below. Impartially it feeds on both their fears and their lusts. Still it keeps rearing, tier on tier, its towers clawing at the sky.

The metropolis, the megapolis, the city—beautiful, corrupt, entrancing, damned—the victory of height over space.

But nature—the mystery of that form of isostatic equilibrium we call nature eventually balances its scales with the mysterious phenomena we call wars, revolutions, catastrophes, plagues, shifts of centers of population, tides of economic change. The very forces that create the city destroy it. Carthage is plowed under. Rome is sacked. Berlin is bombed. London, anemically drained of world power, is a wax image. New York is dying on its feet with the last frenetic convulsions of a chicken with its head cut off. For the day of great cities as we have known them is ending.

The world has shrunk. Superhighways, giant air liners and the radio constrict it into a tiny ball. Helicopters and television will shrink the land still smaller. The great cities of the future will no longer consist of vast teeming populations concentrated in immense stone growths reaching into the skies. They will spread out, their great petals unfolding to light and air, enveloping suburbs miles away. Their tall towers will shrink and vanish. Space at last will have its victory over height.

For years I felt it sad and strange that we had no city in our own vast domain to compare with those I knew in books. The lack was only in my own imagination. We were already citizens of a great city—Pyramid City.

A character named Half-Pint Petey, who had wandered

up for work in the mine, assured me solemnly it was so. He had been everywhere, done everything; he had been a mule skinner, a railroad surveyor on the U.P., a cook for Theodore Roosevelt on his western jaunts, was a good hand with an ax, could turn out blindfolded a pan of breakfast biscuits that shamed Mother's, and swore by *Gulliver's Travels*. A capable and talented man.

Now, son, he said, don't you be grievin' none about never seein' no city. We got one here. A proper world wonder. Naturally it was new. There was plenty of time to get in on the ground floor. This of course was Petey's main obsession. He would propose that we expend the proceeds of the mine for the purchase of an entire valley at two bits an acre; that we secure squatters' rights to a whole plain; establish the water rights to a mountain watershed. Watch her boom then!

Under his eloquent persuasiveness we could see a prairie-dog town metamorphose overnight into a right handsome human town. This, he warned, was but the beginning. Soon all the spaces in-between, all the valleys, plains and parks would be filled. All one vast city extending from the mountains to the river. Pyramid City—the whoppin'est big city in the whole plumb world. Pulling his handle-bar mustaches, Petey would continue.

The Colorado ran right through it. Well, he'd put locks on it like in the Panama Canal. Big ships, by cracky, would sail right up to Pikes Peak and we could buy our tea fresh from China. Another thing, son. We'll have us Moffat Tunnels running through all the mountains. Just ring a bell at a likely ore showing and get off and mine. What's the use of paradin' around on top the hills with a jackass and drivin' separate shafts? And the desert—God Almighty!—put her under glass. Make one big hothouse of her. Grow your own bananas and coconuts. Have parrots squawkin' in the trees, and monkeys climbin' around. It'll be a zoo to

boot, where people can go on Sunday afternoons whizzin'
in the subway.
    Well, we failed him miserably. Half-Pint Petey one day
washed out his extra shirt, stuffed his week's wages in his
tattered trousers and wandered over the range to seek a
better place to begin. But I never forgot his inspired words.
Today there is a paved street where we hunted rabbits, a
corner drugstore on the site of the campfire where we cooked
them. A part of Pyramid City, no less.
    First there's the boroughs, all seven: named Wyoming,
Utah, Colorado, New Mexico, Arizona, Nevada and California.
    Then properly laid out to avoid congestion are the pub-
lic parks: Grand Cañon, Bryce Cañon, Zion, Petrified Forest, Cañon de Chelly, Mesa Verde, Grand Teton, Natural
Bridges and a dozen others.
    Main Street now is labeled Highway 66; it runs east
and west through the southern half of Pyramid City. The
upper half is more full of hills than Rome or Kansas City.
Huge mountains in fact, though residents on one side refer
to those on the other as merely living "over the hill." Central Avenue spans the lot; the Pikes Peak Ocean-to-Ocean
Highway is its present name. Between these two are a thousand cross streets to suit the fancy of every dweller—wide-
paved boulevards, narrow dirt roads, tree-lined lanes. Pick
your house to suit the weather and the neighborhood: log
cabin, adobe or ramada, or a de luxe copy of either.
    Like Budapest, Pyramid City is split lengthwise by a
river. You can't forget the red Colorado any more than you
can the blue Danube. But here we let it sing its own song,
being more orchestral than melodic.
    Geographically our 42nd and Broadway lies exactly in
the center of Pyramid City. Like Times Square, it has its
popular name—Lee's Ferry. For nearly four centuries everybody has eventually showed up here at the confluence of the

Colorado and the Paria. John D. Lee, the Mormon renegade, built its huge ferryboat from pine hauled sixty miles by ox team. Now, since 1929, the Navajo Bridge spans the river. It has the same deathless quality as Brooklyn Bridge; one cannot love one without loving the other. Some 480 feet above water, this is the highest steel-arch bridge in the United States; the only highway bridge crossing the Colorado between Moab, Utah, and Boulder Cañon, a thousand-mile stretch.

Talk about interurbans! Pyramid City's got elevateds, inclines, surface trams, subways and Toonerville Trolleys galore. Its greatest traditions stem from these lines of steel rails. A man could spend a lifetime chasing down the legends and lives of its railroads. The old Colorado Midland elevated that crawled over the roof of the continent. The Mount Manitou and Red Rock inclines that were yanked up the peaks by cables, to say nothing of the Pikes Peak Cog Road with its greasy cogged third rail and an engine on stilts which pushed its single car up the peak. The T.C. & G.B., officially labeled the Tucson, Cornelia and Gila Bend, but better known as the Tough Coming and Worse Going Back. The Denver and Salt Lake—that's a subway for you! Through the Moffat Tunnel, six miles long, it burrows clear through James Peak and the Continental Divide; shortens the distance between Denver and Salt Lake 173 miles; and also brings water through the Rockies from the western slope to the eastern slope. For Toonerville Trolleys we've got a dozen miniatures of trains, complete to detail, high in the Rockies. There's the narrow-gauge lines of the D. & R.G.—"Through the Rockies, Not Over Them"—twisting around the peaks and creeping through the gorges; the little narrow-gauge spur running down from San Luis Valley, lately transported bodily to Burma. And for surface trams the U.P., the San Diego and Arizona, the Santa Fe, the crack fliers with the wonderful names—the Mountaineer, the Colorado Eagle, the Chief, the Navajo, El Capitán, and Grand Cañon Limited. From the historic Santa Fe stems that wholly American legend and incomparable institution which alone in all America compares with the traditional wayside inn of Europe—the Fred Harvey House. No boy who has ever known one but scoffs at a mere diner and hotel.

Spaced mealtimes apart, they spanned our whole wilderness domain. Each was at once an oasis for all travelers and the focus of life for miles around. Not only were they richly named—Castaneda, Alvarado, Escalante, Fray Marcos and Garces after our earliest travelers; El Navajo and Havasu after our first citizens; La Posada and Casa del Desierto as proper names of inns. Their architecture and decoration were distinctly original and disinctive. Nothing phony, scrumptious or Grand Rapids about them. Hung with Navajo blankets, their walls painted with Indian designs, selling Hopi and Pueblo pottery, Navajo silver, and folders of Scenic Views of the Region, they were deeply rooted in our homeland as perhaps no other institution and proud of it—an astounding, isolated fact in the days of their prime.

When The Train Came In!—that was the event for which the Fred Harvey House was expressly built and nobly planned. Before the wheels stopped turning the great brass gong began its deep-toned musical summons. Trainmen and passengers poured off—everyone on board. And everything was ready. In the small dining room hot soup was already steaming on white tablecloths. At the long horseshoe lunch counter of green marble tidy packets of ham sandwiches wrapped in tissue and tied with green paper bands were stacked high, with doughnuts, cakes and pies. Huge polished urns, like the pipes of an organ, played the aromatic tune of Fred Harvey coffee, sweet as the strains of a calliope. Along the walls outside squatted Indian squaws with their pots and baskets and beads spread out on the brick platform. Beyond stood a row of buckboards, the town hack, Studebaker wagons; a line of shouting tourist drivers. And all around milled the townspeople and the countryfolk come in for the mail. . . . When The Train Came In.

In twenty-five minutes to the second it was over. With a last haunting shriek, pouring sand on the rails, the train and all we knew of the World Beyond swung round the curve and vanished from sight.

There was still the Fred Harvey House. On birthdays and gala Sundays we had dinner there, playing we were travelers. At night a cup of coffee, the talk of grimy railroad men, and the rich acrid smell of coal smoke and hot cinders. The Fred Harvey Girls were perpetual attractions. Those in the dining room dressed completely in starched white; those at the lunch counter wore black blouses and black stockings. All were paragons of virtue. They lived upstairs, were up at six and were checked in at ten. The first girl I ever took to a dance was a Fred Harvey Girl. On this occasion she was allowed to remain out till eleven. The romance was short-lived. She married a fireman on the Long Run. Who could compete with such a man? The turnover in all Fred Harvey Girls was terrific, they were in such demand. Always they chose the best—engineers, firemen, brakemen, yardmen, even conductors, all railroad men who neither drank nor smoked, and received the highest and steadiest pay in town.

The Fred Harvey House! At Las Vegas, just below the mouth of our lonely cañon. At Lamy, where the train winds through the pine mountains and the piñons and the patches of bright red adobe. At Albuquerque, down past the old pueblos along the Rio Grande. At Gallup, where you race across the wide, wind-swept plateau. At Winslow, Williams, Ash Fork, Seligman. At Needles, where you cross the river. On the desert at Barstow and San Bernardino. . . . All with their long portales or patios, their great open fireplaces, and their cool clean rooms where you could lie abed at night and listen to the ceaseless prowling of the goat bucking empties along the spurs, and wait for the haunting shriek of the midnight Chief as it swept in out of immeasurable empty space. America's modern mission and only true inn, compounded at once of its rooted past and its ceaseless longing and unrest.

What strange and diverse neighborhoods there are in Pyramid City. Like those of any great city they seem like different towns: Spanish, Anglo, Indian, Mormon, even Chinatowns; foreign quarters, Old Towns, new subdivisions; industrial, agrarian, commercial, arty or merely indolent as Mayfair; yet whose citizenry are linked as closely as any others living round closer corners from one another.

Ouray, snowbound months out of the year in the high San Juans. At night starving herds of elk and deer steal down to eat the hay stacked in the square by townspeople. Lying quietly in bed you can watch their ghostly antlered shapes glide past the window in moonlight.

Gallup, marooned on the high sandy plateaus with more kinds of weather per hour than any other place I know. You can be stuck there any day by wind, rain, dust or snow. A railroad town, a coal-mining town, a center for Indian trading, and one of the toughest towns in the country, it is yet so desolate and lonely that I have heard it has the highest suicide rate in the United States. Stalled drummers, they say, just can't survive a weekend.

Yuma, on the Arizona side of the Colorado, and the old territorial fort prisons crumbling on the bluff overhanging the great bending river. The hottest town in the country. With its sign on the depot lunch room, "Free Meals Every Day the Sun Doesn't Shine."

Bisbee, built on the cliffsides of Mule Pass and Brewery Gulch. The only town of its size in the country where postmen don't deliver letters to the door; the wooden stairs are too high and steep.

Calipatria, Imperial, Brawley, huddling beneath the desert sun of the Colorado Desert in the shadow of their scraggly palms and bougainvillaea vines. The lowest towns in the country.

Such are the strange neighborhoods of our city. The coldest, hottest, highest, lowest, driest, steepest, most desolate and lonely towns in the United States. And precious few at that. What makes them also the most interesting, the most individual, and certainly for many of us, travelers and residents alike, the most loved ones we know? Quién sabe? It is not worth the shrug. But they are the odd street corners, the surveyor stakes between which Pyramid City is building up into one of the great cities of the future. Not into another megapolis, as we have known them, with their dense urban concentrations. But into a new form, with all the solidarity that modern technology implies.

But there is no need to steal Half-Pint Petey's thunder and embroider his analogy further. We can only bow to his prophetic wisdom and say devoutly of his visionary city, "I knew her when."

For in one of its immense and deserted valleys, worth much less than two bits an acre, already had begun work on the superduper waterworks that are to mushroom Pyramid City into the future.

PART THREE

Its Future

CHAPTER I

Imperial Valley

FACE SOUTH and stick out your left hand, palm up. Down the outside of the wrist flows the Colorado River. Your thumb is the Gila coming from across Arizona to make a junction with it at Yuma. Continuing down the forefinger the Colorado empties into the Gulf of California at its tip. All the other fingers are subsidiary channels —a maze of spreading, shifting, floodtime channels.

Gradually the river has built up with its silt a delta bar across the base of the fingers. This ridge of raised calluses is exactly at sea-level elevation, and it forms the United States-Mexican boundary across the upper end of the Lower California peninsula.

Venus Mount at the base of the thumb is heaped with the Algodones Sand Dunes. On the west, up the heel of the hand from the little finger is the crinkly desert range containing Signal, Superstition and Coyote mountains. And on the north, dividing hand and wrist, are the San Bernardino Mountains separating the Colorado Desert from the Mojave.

Within these elevations, and sunk nearly three hundred feet below sea level, is the cup of the palm itself. *La Palma de la Mano de Dios*—the Palm of the Hand of God. As such has this imperial valley in the center of the Colorado Desert long been known.

No palm reader ever found a hand more difficult to read. At the beginning of the twentieth century, fate, heart and life lines of the preceding four centuries had been obscured by driving winds and shifting sand. Across its baking, sterile desert had trudged Melchior Diaz in the sixteenth century and Juan Bautista de Anza in the eighteenth century, followed long after by the emigrant gold seekers of the rush of '49, and still later by the coaches of the Butterfield Overland Trail. But by 1865 the route had been abandoned. All lines of travel had been obscured. By 1900 then the palm was as unlined, bare and empty as it had been four hundred years before.

One of the forty-niners, however, was a man of remarkable imagination: a physician named Dr. Oliver Meredith Wozencraft. Health and finances failing, he had left his wife and three children in New Orleans and plodded west to California. It took him three days and nights to cross the Colorado Desert on a stiff-legged mule. Just enough time to observe the silt in the bed of the ancient lake and to surmise that it lay below the level of the Colorado River.

To be sure of his diagnosis, he returned from San Diego with a surveyor named Ebenezer Hadley. Their investigation confirmed the report of William P. Blake, a geologist with the Williamson expedition, who had taken barometric readings showing the sink to lie below sea level. Wozencraft then worked out a plan to bring water from the river to irrigate the desert basin.

By 1859 he was in a position to promote it, having been selected as a delegate to the Constitutional Convention at Monterey to form the new state of California. First he induced the legislature to pass a bill giving him all state rights to some 1600 square miles of "valueless and horrible desert" in consideration of their later reclamation.

Then he hurried to Washington and worked three years more on a bill to be presented to Congress. It proposed "the introduction of a wholesome supply of fresh water to the Colorado desert." Reclamation was to commence within two years and to be completed within ten. Only then should clear title be granted Wozencraft and his associates. The House Committee on Public Lands gave it a favorable report. But the current California humorist, J. Ross Brown, let loose a clever witticism: "I can see no great obstacle to success except the porous nature of the sand. By removing the sand from the desert, success would be insured at once." Amid a roar of laughter, the bill came up for a hearing in 1862 and was promptly shelved.

Stone-broke, Wozencraft went home and shipped as a surgeon on a Pacific packet for another grubstake. Years later he returned to Washington to plead reconsideration of his bill—"the fantastic folly of an old man." While waiting for it to be introduced again he died at the house of a friend, one of the many men of vision who had lived before his time.

The idea lived on. In 1891 a fly-by-night promoter, John C. Beatty, organized the Arizona and Sonora Land and Irrigation Company to irrigate a stretch of land on the east side of the river. For his technical assistance he hired Charles Robinson Rockwood, an enthusiastic young engineer who had helped to survey the route of the Denver and Rio Grande through the Colorado Rockies, the Southern Pacific through Texas, and the Northern Pacific in Oregon and Washington. After a preliminary survey Rockwood persuaded Beatty to drop the scheme for a better one.

He proposed to irrigate land on the west side of the river in California by tapping the river twelve miles above Yuma at Potholes, carrying the water by a 40-mile canal south and west through Mexico to avoid having to cut through the sand dunes, thence north into the sink. The scheme looked good. Accordingly, the name of the company was changed to the Colorado River Irrigation Company, with the declared purpose of developing land in the Colorado Desert.

But while Beatty was peddling stock the panic of 1893 set in. The company went bankrupt in the faces of its supporters. Rockwood sued for his unpaid salary, and in judgment was awarded all the company's surveying notes, maps and engineering data.

Thus begins one of the most remarkable chapters in that long epic entitled "Winning the West." Wholly, fabulously American, it runs the gamut of skulduggery, bamboozling, stock manipulation, personal vanity, lust for power; and with these, courage, vision, high endeavor, sacrifice and unending work.

In 1896 Rockland formed the California Development Company, incorporated in New Jersey. For two years he combed New York hunting for capitalists to back his scheme. But just as money seemed available, the battleship *Maine* was blown up in the harbor of Havana and war ensued with Spain. An old, grizzled surveyor squatting on the desert near Yuma came to the rescue. Putting up an initial $40,000 and a greater wealth of experience with the river and neighboring Mexican landowners and politicos, W. T. Heffernan kept the new company alive. Then Sam Ferguson and A. H. Heber became interested in the project.

The break came two years later when George M. Chaffey was induced to finance it. Chaffey was a capable man. He was not only wealthy, but a civil engineer and irrigation expert who had developed a large irrigation project in Australia under similar conditions. In 1900 he became president of the company and secured a contract with the Mexican government. Then with $150,000 he undertook the task of carrying 400,000 acre-feet of river water into the heart of the Colorado Desert. Instead of tapping the Colorado at Potholes above Yuma, he connected with it at Pilot Knob opposite Yuma. The main canal was carried south into Mexico, parallel with the river until it reached the barranca, or overflow channel, known as the Alamo, which was then utilized. Then, forty miles west of the Colorado, the water was led north into California again.

Actual digging began on Thanksgiving Day, and on May 14, 1901, the first water was delivered. Immediately things began to happen.

There were then scarcely 1,500 people in the area. And as the California Development Company had been incorporated only as a water-selling company, the Imperial Land Company was now incorporated to lay out townsites and plat the land. Meanwhile a vast advertising campaign was launched to attract colonists. Dread, ominous words like "desert" and "sink" were carefully avoided. Instead, the empty and desolate basin was called by the regal name "Imperial Valley."

Certainly the new settlers found little else to recommend it. With thermometers registering 125 degrees or more, the blinding sun beat down on a desert wasteland of creosote bush, mesquite, barren sand and alkali flats. There was not a tree for a hundred miles, no wood for houses or firewood; only a trickle of muddy red water which had to be strained before it could be drunk. Nevertheless, people streamed in by wagonloads, set up tents and began to break the land.

Stock in the California Development Company shot up on the market. Its original holders, friends of Chaffey, began selling it to outsiders for a quick profit. Rockwood, jealous of Chaffey, raised money and bought control. Chaffey, who had financed the whole development, was driven out, poorer than when he went in, leaving 400 miles of ditches carrying water to irrigate 100,000 acres.

For what? An expert was sent by the Bureau of Soils of the United States Department of Agriculture to make a survey of the soil. He reported it so impregnated with alkali that nothing would grow but date palms, sorghum and sugar beets. It was a bad guess. The rich silt deposits of thousands of years at the first drop of water produced six cuttings of alfalfa a year, long-staple Egyptian cotton yielding better than a bale to the acre, and grapes and melons maturing earlier than had ever been known.

People poured in. There were 7,000 by 1903; 10,000 by 1904. The Southern Pacific built tracks into the valley, bringing still more. Imperial, the first townsite, gave way to Holtville, named after two brother-bankers from Mississippi. In 1903 Brawley became the chief town. As the railroad pushed farther south, still more jumped on the map. El Centro in 1905 was "born of a boxcar" placed on a siding where passengers waited for the junction of the little intervalley "desert sidewinder." Then a new town sprang up on the international border, the half on the Mexican side becoming Mexicali, the half on the California side Calexico. Around them 120,000 acres were soon under cultivation.

It was indeed an Imperial Valley, a rich and heaped Palm of the Hand of God.

The outlaw Colorado meanwhile was bucking at its halter; the first time it had ever been snubbed.

Every single day's supply of water for the valley contained enough silt to build up a levee 20 feet high, 20 feet wide, and a mile long. Within two years, and for four miles below its intake, the Imperial Canal was completely silted up. The California Development Company had no money or machinery for dredging. Crops began to perish, and the farmers threatened $500,000 worth of claims.

At the same time the old familiar battle had been renewed within the company. Anthony K. Heber, a slick and shrewd promoter, had got control as its new president, and he and Rockwood were fighting. To make matters worse, it was found out that the original survey was several points off. Not a settler had clear title to his land. Until this was rectified legal title was still vested in the government. Moreover, the Colorado was still technically a navigable river and Congress had not yet recognized the right of the company to divert any of its waters.

Heber cut this Gordian knot by ordering Rockwood to dynamite the gate at Hanlon's Heading. The engineer was frightened lest he lose control of the river. The president was more afraid of losing control of the company. After a prolonged battle they compromised. In October, 1904, Rockwood cut a new intake in the soft earth four miles below the Mexican boundary. Heber, for his part, entered into negotiations with President Diaz of Mexico for the use of this water diverted on the Mexican side of the line, forwarding plans for a new controlling gate to be approved.

It was a fatal mistake. Approval of the new controlling gate did not come from Mexico for over a year. And, although Rockwood's consent had been based on readings showing that in the past twenty-seven years there had been only three winter floods, a new one rushed down the river. There was no gate to stop it. The water tore through Imperial Canal to Mexico and then north into the valley by way of the old Alamo Channel. The overflow drained into Volcano Lake in Mexico, and this filling up, began cutting another channel, New River, into the valley.

Meanwhile Rockwood had been spending most of his time in Los Angeles contesting Heber's control of the company. In March, 1905, he rushed back for a greater struggle. His first attempt to close the intake by a 60-foot dam of piles, brush and sandbags was a failure. A second crumbled before the wall of water. By June the intake had widened to 160 feet; the river was pouring through at the rate of 90,000 cubic feet per second; and the water was forming a new Salton Sea in the deepest part of the ancient sink.

Heber was frantic. Acting for the California Development Company, he applied to the Southern Pacific for a loan to stop the river on the grounds that the valley was furnishing valuable traffic to the railroad which was endangered.

The request for a decision reached the president of the Southern Pacific, E. H. Harriman. The railroad was deriving some $30,000 in revenue from the valley. The loan needed was $200,000. Harriman made it on the condition that he could name three directors of the California Development Company, one of whom was to be its president.

Heber was forced to assent. He resigned and left for another desert stamping ground, Nevada. The new president appointed by Harriman was Espes Randolph, a civil engineer and former superintendent of the Tucson division of the Southern Pacific under Huntington. After looking over the job he pessimistically wired back to his boss that it might take $750,000 to control the river, and set to work.

Rockwood came out of the reorganization of the company better off than Heber. He was retained as chief engineer, but was given Randolph's assistant to help him—a resourceful engineer named H. T. Cory, formerly professor of engineering in the University of Cincinnati.

Together that fall they made a third attempt to check the river, building a 600-foot barrier dam costing $60,000. On the last day of November there came a flash flood down the Gila. At Yuma the water rose 10 feet in 10 hours, and increased the discharge of the Colorado from 12,000 to 115,000 cubic feet per second. The dam went out like paper.

Here, indeed, was a problem. The Colorado was pouring into the valley at such a rate that Salton Sea already covered 150 square miles. But if the water was shut off completely, 200 square miles of cultivated land would perish. The two men proposed a new plan: to re-excavate the four miles of the original silted channel by means of a powerful steam dredge; and to control the water by a new, mammoth, steel and concrete headgate. The plans were approved and orders let in San Francisco for construction of the great dredge.

They were the plans of mice and men. For that next spring, on April 18, 1906, San Francisco was shaken by its great earthquake and went up in fire. And down the Colorado came the annual spring flood with nothing to check it.

Rockwood's goose was cooked; the day after the earthquake he telegraphed his resignation. Randolph rushed to Harriman in San Francisco. The famous financier sat looking out upon a city wrecked and blackened, almost completely destroyed, then back on his desk to the enormous demands being made upon the Southern Pacific for relief and reconstruction. What he now heard from Randolph was almost as bad.

The late-spring flood of the Colorado was approaching its maximum. The crevasse was now a half mile wide, and through it the river was pouring 6,000,000,000 cubic feet of water every 24 hours. In the middle of Imperial Valley the works of the New Liverpool Salt Company were buried under 60 feet of water, and above them Salton Sea was rising 7 inches higher each day, covering an area of 400 square miles. The twin towns of Calexico and Mexicali were partially destroyed; 12,000 people were in danger of losing their property, homes and lives . . . And yet, by a mammoth jest, there still existed the paradox that these thousands of settlers would be driven out of the desert by lack of water should the river be wholly dammed.

In the ruins of San Francisco, Harriman listened carefully and made up his mind. He would advance another $250,000 to check the Colorado.

Cory, the ex-professor, was appointed chief engineer to replace Rockwood. Early that summer, when the flood of melting snows from the Rockies began to subside, he began assembling men and materials for the task. This time he did not underestimate the power of the river. But it was greater than any man could guess.

That summer alone, as he made his preparations, more changes were made in the delta than during the past three centuries. The new channels formed already had scoured out more than four times as much earth as was removed from the whole Panama Canal. Yet this was only half the danger. The other half lay in the peculiar phenomenon of the "cutting-back" action of the river.

This cutting-back process is easily understood by watching a small stream when it reaches a sudden fall in soft earth. Tending always to equalize its descent, the water quickly eats away the rim of the fall until, by cutting back and filling in, its course spreads out in a long even decline. This is exactly what the Colorado was doing on a vast scale. The fall of the valley from the Mexican line to the gulf was scarcely 100 feet, while that to the Salton Sink was nearly 300 feet. Thus the cutting back of New River was excavating a huge gorge that would continue all the way back to the Alamo Channel, and thence back up the Colorado itself until bedrock was reached. Such a gorge would be so huge and deep all the way from the Salton Sink that all the resources of modern engineering could not divert the Colorado before all of Imperial Valley was completely submerged.

By August, 1906, Cory was ready to begin on the fourth attempt to check the Colorado.

A huge brush mattress 100 feet wide, sewed with galvanized-iron rope to ¾-inch cables, was woven and sunk to the bottom of the river, reaching from shore to shore. A pile trestle was built across the crevasse, and over it were run 300 side-dump railroad cars called "battleships," each with 60 tons of rock. These were dumped on the brush mattress. But to accomplish this simple task the labors of Hercules and fabulous resources were necessary.

All the rock from quarries within a 400-mile radius

was commanded. Few white laborers were available. An attempt to bring 500 peons from Mexico failed. So six tribes of Indians numbering 2,000 workers were mobilized for labor: Pimas, Papagoes, Maricopas and Yumas from Arizona, and Cocopahs and Diegueños from Lower California. Mexico put the whole encampment under martial law, sending a force of rurales under a military commandant.

At last the river was dammed. All waited patiently for the flood to come. It came sooner than expected—on October 11th. Two-thirds of the dam lifted up gracefully and swept off with the tide. Cory watched it go, carrying the headgate that had cost $122,000. Working feverishly night and day for two weeks more on the theory that "putting rock in the brush faster than the rushing current could carry it away was more than a forlorn hope." Cory and his men finally turned back the Colorado on November 4th. Without resting they continued the work until there were dams across both intake and by-pass, connected by levees, forming an unbroken barrier a half mile long. Then on December 7th another flood came down the Gila to support the Colorado's attack. The dam stood. But an earthen levee a half mile south gave way, and again the river began pouring through a new break 1,000 feet wide into the Salton Sink.

This time even Harriman was aghast, having spent a futile $2,000,000 already. Yet there was still another large stake involved. The United States government had started a project which involved the erection of a million-dollar Laguna Dam at Potholes and the reclamation of immense tracts of land. If the river was not stopped now, it would by its cutback action not only fill all of Imperial Valley, but destroy the proposed dam and render valueless all the reclaimable land.

Immediately he telegraphed President Theodore Roosevelt for help in avoiding this national calamity. And now again, but on a greater scale and between greater protagonists, resumed the political, economic and personal battle that had begun with Rockwood and Beatty. For Harriman at the time was being prosecuted by the Interstate Commerce Commission, and President Roosevelt a few weeks before had characterized him as an "undesirable citizen." The president' reply was brief. The United States Reclamation Service could not undertake such a program without the authority of Congress.

And now the wires began to burn as back and forth bickered the president of the United States and the president of the powerful Southern Pacific. Meanwhile the mere humble citizens, helpless pawns as usual, kept on watching their land and homes crumbling away.

The Colorado River would not wait for a Congressional debate. . . . Yet 12,000 people and 1,600,000 acres of government land were in danger of being ruined. . . . The Southern Pacific in assuming control of the California Development Company had assumed also the responsibilty for protecting them. . . . Take it to Congress. . . . But Congress adjourns today for the Christmas holidays. . . .

Finally Harriman assented. Santa Claus had won. The Southern Pacific would proceed, trusting that the government would assist with the "burden." Roosevelt was delighted. When Congress reassembled he would recommend legislation to "make provision for the equitable distribution of the burden."

So just before Christmas the financier and railroad magnate began on what many have called the crowning achievement of his career—to dam 160,000,000 cubic feet of water pouring hourly into the sink.

Two railroad trestles of 90-foot piling were built across the break. Across these were run trains dumping rock into the river faster than it could be swept away. That was all there was to it.

But to achieve this, the Los Angeles and Tucson divisions of the Southern Pacific and 12,000 miles of main-line traffic were tied up for three weeks. Normal shipping from Los Angeles' port, San Pedro, was suspended. Rock was rushed in from the mountains near Patagonia, Arizona, 485 miles away, and from quarries on the Santa Fe and Salt Lake roads. Special trains carrying piling and timbers from New Orleans were given right-of-way. Dumping began. Never before had rock been dumped so fast: 3,000 cars of rock totaling 80,000 cubic yards in 15 days. The whole river was raised bodily 11 feet.

On February 10, 1907, after 52 days of work, the crevasse was closed and the Colorado diverted into its old channel to empty into the gulf. The job had cost over $1,600,000. Amid the rejoicing a bit of news drifted in. Heber had been burned to death in a hotel at Goldfield, Nevada, on the day that the river had been excluded for the first time.

It was time now to take stock. Few men in the world had ever seen such a flood. Again and again Old Man River, America's Mississippi, and China's Sorrow, the Hwang Ho, had inundated great areas, but never for more than a few weeks. Always their waters had drained seaward when the flood was over. But the Salton Sink, deep below sea level, had no outlet. Nothing but the slow process of evaporation could reduce its vast expanse and this was barely enough to offset the normal drainage from the valley's irrigation canals. In the middle of the valley and desert alike it was to remain a great inland sea. All around it stretched ruined patches of reclaimed land; deep arroyos gouged out by the river whose dry, unprotected earth the desert winds would whip over the valley for years to come; buried roads and obliterated railroad tracks; partially destroyed towns; deserted ranches and shacks above which wheeled buzzards and vultures. And buried from sight beneath the inland sea was the New Liverpool Salt Works.

But the Imperial Valley was saved. The bills were footed up.

The Southern Pacific had spent $3,100,000. One million dollars in claims were presented by the Mexican settlers and one-half million by the American settlers of the valley. The New Liverpool Salt Works added another half million. With still another one million in odd claims, the total costs amounted to some $6,000,000.

How much of the Southern Pacific's share the government would bear in accordance with the president's promise to "make provision for the equitable distribution of the burden" remained to be seen. On January 12, 1907, in the midst of the fight, Theodore Roosevelt had vigorously presented the issue to Congress. The following year a bill to reimburse the Southern Pacific was introduced in the House. President Roosevelt gave it his full support as he had promised, and it was backed up by the public at large. Yet no action was taken.

In 1909 Roosevelt was succeeded by Taft. In his first message to Congress the new president recommended payment. For two years more nothing was done. Finally in 1911 the Committee on Claims considered the matter. First the members reduced the actual amount of $1,663,000 spent by the Southern Pacific on the last successful attempt to $773,000, discounting all the previous attempts. Then they reported it to Congress with the recommendation that it be passed. Five members, however, formed a minority to block it. Their report described the bill as an "attempted raid on the Federal Treasury," denied "any legal, equitable or moral obligation on the part of the Government" to pay it, and referred to the proposed sum as a "gift of the people's money" ... A gift that has been calculated to have saved the people within five years a sum equal to the interest on a $500,000,000 investment!

Thus ended the matter. But for years afterward the

controversy went on as to the right of the government thus to refuse payment morally if not legally promised by its duly elected chief executive, and as to the exact size of the laurel wreath that still remained to be fitted on the head of that "undesirable citizen," the great public benefactor, E. H. Harriman. There is no doubt that Harriman's unselfish action and initiative are wholly commendable. It saved the valley. For, unlike the farmers under United States Reclamation Service projects who were carried by the government during bad years, the settlers of Imperial Valley never had received help from their government. A million and a half dollars, even three million, was a great deal of money in 1907, and it was nobly spent from Southern Pacific funds. Funds, however, that were derived from similar ventures, and which as a long-term investment have probably proved most profitable. And at the same time one must wonder how much, if not more, Huntington had spent before Harriman on that vast estate and enormous collection of European paintings and manuscripts, now the famous Huntington Gallery in Pasadena.

It is a significant controversy in still another respect. The utilization and damming of the Colorado threw into early focus the basic differences between proponents of government control and advocates of private enterprise for the development of such great natural resources. All the ramifications of the gigantic struggle were there: the initiative and risk of development, the final responsibility, and the rights of the people themselves.

A few years later the valley was threatened again. Building up its bottom with silt as before, the river broke through another breach a few miles south of the dam. Cut off by levees from the Alamo and New River channels into Imperial Valley, the flood poured into Volcano Lake south of the border. Here it deposited its silt and drained south, this time into the gulf by way of the Hardy and Pescadero channels. Hence a new levee was necessary from Hanlon's Heading all the way to Volcano Lake and thence west to the Cocopahs. The river still continued rising higher each year. So in 1922 the Pescadero Cut was made, diverting the river from Volcano Lake into the Pescadero, a tributary of the Hardy that by now had become the main channel. This at best was but another expedient.

The Colorado River, confined for so many hundreds of miles within deep gorges, was here an "overhead river" traveling on a bed from 100 to 300 feet high. Below it lay the rich delta of America's Nile, Imperial Valley, La Palma de la Mano de Dios, America's Winter Garden. It contained over 60,000 people, property values totaling $137,000,000, and nearly 750,000 acres of irrigable land. All the valley was dependent upon the whim of the Colorado; threatened in June, July and August with permanent inundation, and suffering from drought in September and October. It was the last frontier in the vast wilderness basin of the Colorado.

Such was the Imperial Valley when I was sent there in 1925. It was not much older than I was, and we were both romantic.

The valley's romance was portrayed by the murals on the lobby walls of the Barbara Worth Hotel. They pictured scenes from *The Winning of Barbara Worth* by Harold Bell Wright. Even the hotel stationery carried a cut of the fetching young lady herself under a great romantic sombrero whose wide brim was coyly upturned. For her this huge, terrible desert had been conquered by brave engineers and opened to courageous pioneers. Not a hint of Beatty and Heber, of stock manipulation and quarreling bankers stained the walls with a splotch of perfidious truth and crass reality. The hotel was the only modern hostelry in the desert, and it was air-cooled to boot. Rooms cost a mere $5 daily and meals served from silver stamped with Miss Worth's face and hat $2 more.

My romance derived from the fact that I did not have to pay a cent of this modest stipend; it was all covered by company vouchers. For I too was a brave, though junior engineer and this was my first real job. For a day laborer just come from a pipe-line gang in the Salt Creek oil fields of Wyoming, this was romance enough.

Everything tied in to the illusion. Rockwood had just died after being reinstated in the Imperial Irrigation Company, which had succeeded his California Development Company. Nothing else had changed. It was May. The first of 14,000 carloads of cantaloupes were being shipped. Eastern brokers thronged the lobby, telephoning Kansas City, Chicago and New York to order diversions of refrigerator trains as the fruit market rose and fell with the barometer. Ranchers in heavy boots stomped into luncheon conferences demanding water, railroad cars, trucks, labor, and stayed to bet their profits at poker that night in the back room. Outside lay the vast green oasis of the valley producing $100,000,000 worth of its world-famous melons, cotton, vegetables, grain and fruits.

All the little towns bustled with activity—El Centro, Imperial, Brawley, Holtville, Calexico, even Heber Beach, Westmoreland and Calipatria, still clumps of wooden shacks and shipping sheds. In each, long portales like shadowy corridors of Spanish mission churches protected sidewalk travelers from the blazing sun. Between them ran the rough and rutted desert roads perpetually beclouded with dust as fine as talcum. Tiny elf owls dotted the tops of fence posts; roadrunners darted from the shade of smoke trees and tamarisk; lizards baked in the sun. Beyond on all sides shimmered the bare and tawny desert mountains. And over all, land and towns and people, the glamorously sinister Harold Bell Wright desert cast its spell.

My boss was comfortably located in San Diego. Like all lovable gentlemen of girth he eschewed the summer heat and never set foot in the valley. "Why in the hell should I?" he would explode by long-distance. "That's what you're

being paid for. Suppose you bring over your reports and let me see if you've been doing anything more than keeping out of the sun." So at four o'clock of an afternon I would ride a train on the newly completed San Diego and Arizona Railway line up and out of that great cupped palm. Usually I was the only passenger on the observation Pullman. Loneliness but enhanced the view.

One sat dreamily watching the vast rock-rimmed bowl deepen and fill with a violet haze as the train puffed slowly up the steep grade. Great cañons of baked brown rock fell away on both sides. Little clumps of native palms darkened their deep floors. Steadily now we bored through the mountains—through 21 tunnels in 150 miles. Carrizo Gorge, too little known in America, is named for the grass in its depths, which is used by Indians in basketwork. It swallowed us in its magnificent, 11-mile long and 1,000-foot deep immensity. At twilight George Washington Smith appeared in his starched white coat. "Hungry sir?" I was the only diner, it appeared. Why bother with keeping the dining car open? A tray inside, anytime, would do as well. "Yes, sir!" Now this Special Meal For One—"What do you suggest, George?" With a straight face and in university English he would reply. "Coyote steak pan-fried in the fat of a Gila monster's tail, sir? Perhaps a fricasseed Arizona road-runner garnished with creosote bush. Or may I suggest a cool salad . . . of prickly pear cactus with a dressing of river water and . . ." It was a little game we would play week after week. And faster now, through pines, great live oaks and cool fresh air, the train rushed on through the night.

On one of these trips we ran head-on into reality. I had got hold of a copy of Harold Bell Wright's masterpiece.

"The desert waited, silent and hot and fierce in its desolation, holding its treasures under the seal of death against the coming of the strong ones," I began to read.

The Strong Ones came, and the Strongest of the Strong. He came and conquered, beating the desert to its knees. And finally on the last page I read the meaning of the desert that he had won with victory. That great secret of man's relationship to the living earth that he had conquered, as interpreted for us by one of the most renowned American writers of his generation.

"Barbara," he cried. . . . "Your Desert has taught me many things, dear, but nothing so great as this—that I want you and that nothing else matters. I want you for my wife."

Next morning my boss, under instructions from the Big Boss, informed me that I had been cushioned on company expenses long enough and had better settle down for a long stay in a place of my own. One finally became available. I moved out of Miss Worth's namesake into a two-room shack costing the exorbitant sum of $40 every month.

The plank walls were waist high, to the top of which canvas flaps could be let down from the roof in daytime to shut out the sun and rolled up at night for air. At 125 degrees Fahrenheit the place when closed was an oven. Opened to the breeze, bed and floor were quickly covered with sand. The greasy kitchen stove and sink swarmed with cockroaches large and hard-backed as turtles. Regularly the town's water-settling plant went out of whack, and then from the faucet the shower contraption poured the muddy red Colorado itself. One took a bath and brushed off with a whiskbroom.

Dreading bed, people lay all evening on the parched grass lawn of the tiny park. Swarms of moths fluttered round the street lamps. Under this flickering light a tall young lawyer, formerly employed by the Imperial Irrigation District and now a congressman, harangued the crowd. His name was Phil Swing. With a fellow named Johnson he was writing a bill to present to Congress. It proposed a great dam in the upper cañons of the Colorado to save the valley from destruction. Also an All-American canal of some sort. What did the folks think of it? Everybody clapped of course. But really it was too hot to bother with thinking.

One listlessly strolled back through town. A drove of ground crickets had swarmed in from the desert; the downtown streets were black. So one walked the lonely off-side streets, the cottony desert streets of night. It was like walking through a hospital ward. On the withered lawns, under the papery palms, row upon row of figures lay sleeping on their blankets.

But there was always Mexicali, a half hour south down Mexico way. *Mexicali Rose, I'm dreaming . . . Mexicali Rose, I love you . . .*

A little adobe village strung along the deep dry arroyo of Rio Nuevo, it was at once the mud-walled capital of the northern district of Lower California and the dusty center of the Mexican half of the valley, a Barbary Coast in the Colorado Desert and a Sink of Iniquity within the Salton Sink.

It embraced all the features of the boom mining camp of Cripple Creek high in the Colorado Rockies; the squatter camps of Teapot and Lavoye in the Salt Creek oil fields of Wyoming; and the toughest railroad town in America— Gallup. This was its heyday. It recapitulated the frenzied boom days of every mining camp, every oil camp, cattle town and rail end throughout the whole Colorado River basin. And it was the last. Built on the greatest treasure of all—the land itself. Rich delta soil that boasted better cotton than the deltas of the Mississippi and the Nile, the finest melons on earth; that produced a crop for every month of the year. Water so potent it would grow hair on a billiard ball. Mexicali was the newest and the last frontier town in the basin. It was riproaring, hell-firing. All to the tinkle of guitars, to the musical tone of soft Spanish. *Borrachita, me voy* . . .

As long as I shall remember, the great blazing signs of the "Southern Club" and the "Owl" will loom aloft in the velvety night as portals of a world that will never exist again. A perpetual carnival of guitars, slot machines and rickety player pianos you could hear a mile away. The hot blast of an alcoholic stench, and that faint indescribable odor of acrid dust and urine which is back-street Mexico, you could smell a mile farther. Down its wide dusty street lurched Mexican labradores, Negro cotton pickers, giant Hindus with heads swathed in huge black turbans, a thousand shuffling Chinese, and crowds of Prohibition-thirsty American ranchers who swarmed across the border nightly.

Forty cantinas, bars, casinos, cabarets and drinking dens opened to receive them. All adobe, with pink, green, purple, yellow and red fronts—El Palacio, Tivoli, the Black Cat,

Casa Blanca, Mezcal, the King of Cotton, River's Mouth, the Blue Fox, Oriental. . . . In the best, things ran high, wide and handsome. Revues and floor shows every hour on the hour. Cutlets of venison killed off-season in the sierras, quail from the chaparral, fish from the gulf. Served with champagne and the finest wines from Portugal and Spain, duty free. The best of Chinese food. Behind in the gaming rooms tables were reserved for the elite. Not a vulgar greenback, not a piece of silver was allowed to cross the felt. Only gold. Heavy, yellow gold. My friend and later roommate, Ricardo, presided over the largest. Part Mexican, part Indian, his dark, tragic half-breed face looked as if it were cast of bronze. He wore the finest pongee shirts; his long pendent eardrops were of a clear turquoise. He worked four hours a day, received $20 a day straight salary, and neither smoked nor drank. Behind him at the bar stood George, a bartender, my friend of ten years, the most discriminating literary man I have ever met. And down the room were table after table of poker, both pocar robado and pocar garanona, blackjack, écarté, pachisi, fantan, dice and chuck-a-luck. One drank fast, gambled hard, with an eye on the clock. For at nine o'clock the border closed.

At five minutes to nine began the exodus back into the United States. Night after night they poured through the great wire gates—a passing parade, the finale of the greatest show on earth. Then the whistle blew. The gates clanged shut. The lights of the great, gaudy cabarets went out. There was only a dreary little Mexican village swathed in the flocculent desert night.

Now in the flicker of ocote torches at the street turnings, drab little loncherias offered tortillas, tacos, greasy plates of beans and meat. By lamp- and candlelight dreary mud-floored drinking dens selling mezcal and green tequilla began to fill. Chinese lottery vendors took up their stands. The ill-lit bars became alive, and the cribs behind them—

large open courtyards, Mexican style, holding twenty to forty girls each.

The underworld, moving slowly and cautiously like a snake, began to circulate. The professional gamblers, the dealers and the bartenders. The hundreds of half-naked prostitutes in their scuffed, colored slippers and sweat-splotched shifts. The thousand sateen-coated, sibilant Chinese. The horrible beggars crawling from the gutters with the ground crickets and the cockroaches. The brutal-faced mestizos, chollos, criollos, coyotes and cross-breeds. All the rateros, the pimps, the petty criminals and refugees of both countries, all the drunk and the dissolute, the marihuana addicts and the hop-heads, the damned and diseased of this greatest slum in the whole basin—all those who by their labor and the prostitution of their lives forever lay the invisible foundation of progress, and make of each frontier the strangest, cruelest, most pitiful and most alive spot on earth.

For a boy in his early twenties such a spot has an inexorable, fatal fascination. Yet nothing in his surroundings seems strange; all that presses him from all sides seems familiar. For its pattern is one that he has known from childhood, its substance has created the stuff of his own life. And as he lurches drunkenly along a now hushed and fetid lane at midnight under a wan desert moon, all that he has just known and seen and smelled and sensed seems to him more rich and powerful and naked and vibrantly alive than anything he has ever known and loved. A teaming life fecundant as the virgin earth from which it sprang. A new continent not yet seeded to the future, a new life that not yet has borne fruit.

He stops suddenly. He had stumbled over the lap of a prostitute fallen in the doorway of a dark cantina. Pobrecita! Poor drunken pajarita! He stoops and touches her on the naked shoulder. Without a sound she falls over on her back. Sticking out from her flabby, discolored breast is the bone handle of a knife.

He is not frightened, for his own youthful innocence has clothed him in an armor invulnerable against harm. He is too drunk to feel horror, pity, anger, responsibility, or even to listen to that philosophy of acceptance which would have told him that in the tide of life perpetually swelling against mankind's frontiers death means little. So he merely staggers on to the belated and debilitated clatter of a far-off player piano.

Mexicali Rose, I'm dreaming . . . Mexicali Rose, good-bye . . .

The sister border towns of Mexicali and Calexico, whose names were composed of syllables from "Mexico" and "California," were really one town whose halves were separated by a high wire fence. And it was but one of a dozen similar towns strung upon the international boundary from San Diego to El Paso.

Here, along a thousand-mile stretch, it was as if two great waves of opposing culture had met and gouged out a deep chasm, a dim twilight world, between two races. Like great remaining pools these border towns commingled the jetsam and flotsam of both. Neither wholly American nor wholly Mexican, they had a peculiar character of their own. When I knew them most intimately in the roaring twenties, they seemed perpetually hopped up as if on marihuana. Border Town then, wherever it was, was a screaming oasis for Prohibition-thirsty Americans, a bedlam of bars, bagnios and casinos.

Tijuana, south of San Diego, was in the height of its fame. It boasted the longest bar in the world and fifty shorter ones. The San Francisco, where George worked for a time, was a favorite. In the Cerro Azul, the "Masked Woman" disrobed nightly. We liked Caesar's Place best, a chummy little dive that held scarcely a dozen people at a time. The two Italian proprietors claimed that they had brought from the Barbary Coast the secret of the famous Pisco Punch, and allowed only two to a customer. On the profits each of the brothers soon opened a big house of his own.

But just outside of town was built a resort for the more fastidious. Agua Caliente was a gorgeous palace Spanish in architecture, American in management, French in cuisine, and Mexican in service. Costing nearly six million dollars, it had beautiful long portales, cool patios, private cottages, dining rooms that fed two thousand people daily, gambling rooms, and for the main attraction a race track that replaced the old Coffroth track in town.

Tecate, high in the mountains of chaparral and live oaks to the east, sprang up. Little more than a village, it owed its importance to the road being cut through the mountains from Mexicali to Tijuana. Over this groaned trucks loaded with beer, saving transportation through the United States.

Algodones, meaning "cotton" in Spanish, is no longer to be found. It lay just south of the border, west of Yuma, a dreary and sinister huddle of shacks baking in the desert. Only at night it came alive with all its cantinas: El Rey del Algodon, Supremo Mezcal, La Luz Colorado, Esmeralda, Chicago, Lone Eagle Cabaret and El Nuevo Mundo. It was here, I recall, that I was offered my most novel business opportunity—partnership in a crib of forty girls.

Halfway across Arizona lay another pair of twin border towns, Nogales, Arizona, and Nogales, Sonora. The new train route to Mexico down the West Coast had just been opened; Nogales was its gateway, and celebrated the event with a prolonged fiesta. George had moved there to become head bartender in the Cave, the newest and finest cabaret of its period. Formerly the Mexican jail, it was dug into the tall cliffs flanking the town. Each cell was now a private dining room.

Nogales seemed then like a movie set. Resplendent Mexican generals stalked through the bars with all the castes of the American sporting fraternity; shiny Packard limousines bumped over the streets behind plodding burros loaded with wood; shop windows gleaming with jewels stood beside great open markets heaped with enormous red pomegranates, aquacates, little green tropical oranges, lemons, limes.

Then suddenly a new reel unwound. The resplendent Mexican generals became more pompous; three thousand soldiers swarmed the narrow streets; long trains of troops rushed southward. Another Yaqui uprising was expected at Magdalena, just south. Here on October 4th each year all the Yaquis gathered from the wild Sonora hills, a pilgrimage essential to them as the journey of Moslems to Mecca. Two centuries ago the Indians had been carrying the image of St. Francis north on a palanquin when the saint got tired and desired to go no farther. So they set him down and built on the spot the church of St. Francis, still the patron saint of the Yaquis. Here in this church the priests believed the body of Padre Kino was buried, instead of at San Ignacio. But hardly had they attempted to substantiate their claims than Governor Rudolfo Elias Calles issued, on September 29th, an edict expelling all Catholic priests from Sonora, and ordered the mission at Magdalena closed.

It was now October 3rd. The image was reported to have been taken from the church. Thousands of angry Yaquis had gathered, demanding that St. Francis be returned to his pedestal before midnight. So now, on a crowded troop train with all lights out, we rushed southward through the night. How beautiful it was, the little village in the river's bend, the spire of the church protruding in the dawn! People flooded ancient Magdalena: Yaquis camped in field and plain, in the very streets; soldiers helpless to hold back the crowds; villagers barricaded in their stout adobes. Then suddenly at dawn the rush—a slow but irresistible mass of humanity edging forward inch by inch to the church. We could see crushed figures being brought out, a trampled woman, a soldado with his uniform torn to rags. There was a splintering roar; the huge carved doors gave way. Then a mighty sigh, a great quiet breath of relief. The very silence changed tone. It was no longer ominous. It was only patient. I saw, far behind me, a Mexican general wheel his horse toward the hotel for breakfast. The Yaqui uprising had been quelled. St. Francis still sat on his ancient pedestal . . .

Bisbee and Naco were another pair of border towns farther east. Bisbee, on the American side, clung to the side of the mountain; a huge, open copper pit, a beautiful old hotel—the Copper Queen, and hundreds of tiny miners' shacks and houses lining the steep sides of Brewery Gulch, once the most famous red-light district in the Southwest. Far down below it on the plain lay Naco, sleepily staring at the San Jose and Mule mountains in the distance.

Farther east on the Arizona line were Douglas and Agua Prieta, alone and lost in space. I remember them best for a bearded stranger with whom I struck up an acquaintance over a bottle of beer. Jones or Smith or Jackson—whatever his name was—was a collector of snakes for various museums, and had made several trips through Mongolia. He was here for a vacation. It had a mighty purpose. He intended to prove that the legendary feathered serpent of the Aztecs, deified as the god Quetzalcoatl, existed in fact.

Yes, sir! A queer species of rattlesnake with small protuberances that undoubtedly had been wings. They existed in the high sierras of Mexico between Sonora and Chihuahua, and he was outfitting an expedition to go after them. In a few days it left: Jackson and a couple of Indian guides followed by a string of half a dozen pack animals. For several weeks it received considerable press as it proceeded through several villages. There was a long hiatus. Then again the newspapers picked up the story. Jackson, long overdue, had not been heard of; Mexican authorities could find no trace of him. Jackson had vanished completely in the wilderness and, as far as I know, has not been heard of yet.

Still farther east, near the mid-point of New Mexico where Texas cuts in, were El Paso and Juarez, the last of the border towns along our stretch of border.

It is amusing to recall the controversies of the period. The Mexican federal authorities requested that all ports along the international border be kept open open for twenty-four hours. Immediately the American Associated Chambers of Commerce and various city councils protested to President Hoover. Lengthy petitions urged that the border gates continue to be closed at 9:00 P.M. unless all gambling be stopped, as gambling "constitutes a great economic drain upon the community."

Imperial Valley citizens beseeched the Mexican gambling houses to substitute blackjack for écarté; the latter game, they said, was too tough to play. Still another problem came up when Mexican officials decreed that all persons working below the line except "technicians" and "specialists" must live in Mexican towns. In Calexico more than three hundred persons were affected. American card dealers and bartenders set up a howl of consternation, then overnight became "technicians" and "specialists."

While these controversies raged, line-running of alien Chinese and opium was giving serious trouble. Chinese were slipped across the line by every means conceivable—boxed in crates of melons, disguised as old Mexican señoras, and even carried by plane from Laguna Salada. The usual route was from Mexicali's Chinatown, a sinister quarter of narrow, crooked alleys, to the Chinatown of Los Angeles, now replaced by the new Union Station, and thence to San Francisco. Here for a brief period of six or eight years were reenacted bloody melodramas that followed the traditional pattern of those of the 1890 Chinatowns.

The comic opera of the border was the Mexican revolution of 1929. Every border town played its part. The revolution got under way slowly; the rebel generals had to be outfitted in proper style. In Naco a fancy hat dealer sold 500 hats, some of which cost $60 apiece. One general, he said, bought nine hats and nineteen silk shirts in order to revolt with sartorial elegance. How handsome they were, rebels and federals alike! General Yucopicio, commanding the garrison at Agua Prieta, was a Mayo Indian—a beautiful figure of a man, self-contained and impressive.

But by spring everything was ready. The revolutionists, commanded by General José Gonzales Escobar were marching on Nogales, Naco, and Juarez. Sweeping up from Zacetecas were the federal forces commanded by General Andrea Almazan under the direction of General Plutarco Elias Calles, secretary of war. Lower California was still loyal. General Abelardo Rodriquez was reported marching from Mexicali to San Luis, just below Yuma, to combat rebel troops under Governor Topete of Sonora. A wonderful exhilaration, a tense excitement, pervaded the whole border. There were manifestos and countermanifestos; American newspaper reporters thick as fleas; hordes of ragged soldados, rancheros and paisanos pouring into every town. History was in the making.

On March 8th this great revolution, like an immense firecracker, was touched off.

That night in Mexicali bombs began exploding, and pandemonium broke loose as inhabitants jumped from bed to scurry for shelter from the ghastly surprise attack. Aye de mi! A short circuit in the motor of an airplane housed in the government hangar had touched off some bombs and a fire. Señora Fuentes, wife of the chief mechanic, was slightly injured.

That same dawn the rebels attacked Juarez, swarming into the main streets, Calle Commercio and Sixteenth of September. Shouts from the open bars and gambling houses, a patter of gunfire! By noon it was all over. The city had fallen, and Teddy Barnes, bartender of the Mint Café, had been killed by a stray bullet.

Nogales and Naco fell next. How charming were all these stirring dramas! The resplendent Mexican generales. The cavalry on beautiful horses riding in cowboy saddles with stirrups so long the toes just touched. The violent denunciations, the patriotic fervor. Suddenly an outbreak of gunfire, a token resistance to preserve honor. Then the soldiers pouring into the streets—long ragged lines of Yaqui warriors, shoes thrown away, and plodding in old huaraches with feet blackened and hardened to the consistency of jerked beef. Next came the women with their charcoal braziers and cooking pots—the mothers, wives, sweethearts and camp followers.

Viva Mexico! Bonfires and campfires lighted the streets. The bars and bagnios threw open their doors. Except for red armbands on the rebels, you could not tell them from the federals or the villagers. They were all one people, intensely alive to life, ignorant and uncaring of theories. Innocent children fighting a mock war without bloodshed, without bitterness, and thus maintaining a racial integrity that endeared them to us all.

Viva! Viva Mexico!

CHAPTER 2

The Colorado River Project

FROM TIME immemorial the Colorado and its tributaries have been used for irrigation. A jar of water carried up a steep trail to a corn patch on the shelf of a cliff; a handful of melon seeds thrust into a flood bank of rich silt to be watered by the overflow of the river—thus did the prehistoric pueblo dwellers first utilize the river, as the Havasupais do today in the depths of the Grand Cañon. The Hohokam of Arizona, the ancient Canal Builders, constructed extensive irrigation works. Los Muertos, a city of thirty-six communal buildings antedating Columbus' discovery of the New World, was supplied with water from the Gila River by a large canal 30 feet wide and 7 feet deep, with a network of side canals for irrigation. Perhaps no better irrigation engineers existed than the first white settlers in the basin—the Mormons. Throughout all Utah are remains of their earliest diversion dams, canals, irrigation and drainage ditches, all built and used co-operatively as community enterprises.

Imperial Valley was the first large-scale, private irrigation project to utilize the Colorado itself for reclaiming an uninhabited desert. A few years later another private colonization plan was promoted at Palo Verde just upriver. An immediate success, it opened up nearly 70,000 acres and gave birth to Blythe, a town of 2,000 population.

317

THE COLORADO

Swiftly now there followed a series of governmental reclamation projects developed in an effort to utilize other sections of land. The Yuma Project, costing nearly $9,000,000, made available 110,000 acres for irrigation. It included the construction of the Laguna diversion dam which permitted water to be taken into the California side of the Colorado, then by means of an inverted siphon to be carried under the river to the Yuma Valley, benefiting more than 6,000 people. The Salt River Project in Arizona, comprising the Granite Reef dam and the Roosevelt storage dam, opened 213,000 acres to irrigation at a cost of over $10,000,000 and made possible the mushroom growth of Phoenix, the only city in the Colorado River basin of 50,000 population.

In the upper basin three other projects were developed. The Grand Valley Project, near the convergence of the Colorado and the Gunnison, including the erection of a diversion dam, made possible the irrigation of 55,000 acres and the growth of Grand Junction, Colorado, at a cost of $4,000,000. The Uncompahgre Project, also in Colorado, provided a 3-million-dollar tunnel six miles long and running one-half mile under the mesa to bring water from the Gunnison River to irrigate 97,000 acres in Uncompahgre Valley. The Strawberry Valley Project, Utah, built on the Mormon tradition, was no less novel. An earth-fill storage dam and reservoir was built to provide for the storage of water in that part of Utah within the Colorado River drainage basin. Water from this was then led by the Strawberry Tunnel four miles through the divide into the Great Basin of Utah, utilizing in part the channel of the Diamond Fork of the Spanish Fork River. At a total cost of nearly $4,000,000, over 55,000 acres and 16,000 people were benefited.

Nothing illustrates more clearly than these early projects the character of the basin and the difficulty of irrigation. To utilize even a small portion of the water in isolated spots

COLORADO RIVER PROJECT 319

diverse and novel means were required. The cost was enormous. Comparatively small areas and few people were benefited. The population of the whole seven-state watershed area of 180,000 square miles above Parker was but 315,000 —an average of 1.7 per square mile.

Yet these few projects were minor samples of what eventually would have to be done. The basin was still an immense wilderness—hundreds of small mountain valleys, great sweeps of barren plateau and vast expanses of desert only waiting for water to make them productive as any fields and orchards in the world. And sweeping savagely through it a river of precious water, two-thirds of its total annual volume pouring unused into the sea.

Only against this background does the Colorado River Project begin to loom in true perspective.

Like all big concepts it grew slowly.

In January, 1919, a conference was called in Salt Lake City by the governor of Utah to discuss the use of the water of the Colorado River. Attending were representatives of the seven states affected: Wyoming, Utah, Colorado, New Mexico, Nevada, Arizona and California.

The governor put their case neatly: "The water should first be captured and used while it is young, for then it can be recaptured as it returns from the performance of its duties and thus be used over and over again."

The conference arranged for the formation of the "League of the Southwest," which met twice the following year, at Los Angeles and at Denver. It was suggested that a compact be entered into by the seven states and the United States government. Accordingly, in May, 1921, the representatives of the league met again in Denver and formulated resolutions to be laid before the president of the United States, the secretary of the interior, and Congress.

The federal government was as interested as the states concerned, and named Secretary of Commerce Herbert Hoover to represent it. He and the seven commissioners then met at Santa Fe, New Mexico, to negotiate the Colorado River Compact, which was signed on November 24, 1922. The compact divided the waters of the river into two great drainage areas separated at the head of Grand Cañon. The Upper Basin was composed of Wyoming, Utah, Colorado and New Mexico; the Lower Basin of Nevada, Arizona and California. Based on the annual total volume of 20,500,000 acre-feet, 7,500,000 acre-feet of water were apportioned annually to each basin, and the Lower Basin was given the right to use an additional million acre-feet annually. It was further agreed that if any river water were allocated to Mexico by an international treaty, it would be supplied first from the surplus above this total of 16,000,000 acre-feet, and if this were insufficient the deficiency would be borne equally by the two basins.

No attempt to apportion the water among the several states was made; only between the groups of states comprising the Upper and Lower Basins. Furthermore, execution of the compact was subject to ratification by all seven states. Yet Arizona refused to ratify it.

Meanwhile another aspect of the river had come up for consideration. The young Imperial Valley lawyer, Phil Swing, had gone to Washington in 1919 to present the flood danger of the "overhead river." He pointed out that there were but three depressions along the delta bar separating Imperial Valley from the gulf through which the Colorado could empty into the sea. In 1906 one of these had been filled with silt at the time of the great flood. By 1922 the second was to be filled. Already the third depression was filling. Unless a fabulous sum was spent in Mexico to dredge an artificial channel, the river was certain to pour into Imperial Valley.

Paradoxically, at the same time the valley faced ruina-

COLORADO RIVER PROJECT 321

tion by failure of its water supply. The canal diverting water from the river ran sixty miles through Mexico, and more and more of the water was being used for the great cottonfields of the Mexicali district.

Swing, with two other proponents, Rose and Kettner, proposed building a new All-American canal to be run wholly through American soil to irrigate the valley. It was to be paid for by Imperial Valley on a loan from the United States. Elected to Congress the following year, Swing redoubled his efforts. Congress was persuaded to pass the Kinkaid Act, directing an investigation of the lower Colorado. At the same time the Reclamation Service began considering a major governmental project in one of the upper cañons.

Immediately a strong opposition set in, headed by a small group of American capitalists who owned land in Mexico. Chief of these was the owner of the powerful Los Angeles *Times,* Harry Chandler, a bitter conservative and reactionary financier whose fingers were in every development pie in Southern California. Controlling tens of thousands of acres of cotton in the Mexicali district, he was naturally opposed to any All-American irrigation project that might withhold water from his Mexican holdings. Using the *Times* as his mouthpiece, Chandler began an obstructionist campaign against either a dam that might stop the free flow of the river or a canal that might divert its waters.

Luckily William Hearst, just as reactionary, happened to be his press rival. To combat Chandler he came out for both dam and canal. The public began to be aroused. A group of citizens met in California and formed the "Boulder Dam Association" to advance the construction of dam and canal and other river development projects in the Lower Basin states. Membership consisted of farm bureaus, county boards of supervisors, chambers of commerce, and various other civic groups—over 200 such organizations in Southern California, Arizona and Nevada. They began to carry on a campaign of public education through the medium of newsletters, pamphlets and booklets.

The fight had begun. For eight years it dragged on in Congressional anterooms, state and federal offices, hotel lobbies, press editorial rooms, luncheon clubs, and on street corners.

Meanwhile another fight had started between a small group of quiet men and the immense forces of the river itself. Month after month surveyors, geologists, hydraulic, construction and reclamation engineers were plodding up and down the river studying its cañons and submitting their reports to F. E. Weymouth, chief engineer of the United States Bureau of Reclamation.

Glen Cañon was first proposed as the proper site for the first major development. Investigation showed that the soft limestone walls would not support a great structure; the cañon walls were farther apart than those of Black or Boulder Cañon downriver and hence would require greater yardage for a dam; and there were no superior storage facilities behind it if a dam was built.

Bridge Cañon, superficially an excellent damsite, was surveyed. Its storage possibilities were limited and would have to be supplemented.

Next proposed was a low flood-control dam at Topock —the Mojave or Needles dam. The plan provided for its construction either at direct government expense or as a charge against the lands to be protected. This too was given up. It would cost as much as a storage dam at Black Cañon, destroy a large area of reclaimable land and being of no use for the development of power it could not reimburse the government for its cost.

After the investigation of eight damsites, the choice finally narrowed to Boulder Cañon or Black Cañon just below it. Here the cañon walls ran 1,000 feet above the river, sank 150 feet below the low-water level, and were but 1,000

COLORADO RIVER PROJECT 323

feet apart at the crest of the proposed dam. It was close to Las Vegas, Nevada, for construction purposes; close enough to Los Angeles, California, to supply a sale of power; and it offered the required storage facilities. But a stupendous dam would have to be built—one that could stand a stress of 30 tons pressure per square foot. It might cost $72,000,000. And if it should fail, "the flood created would probably destroy Needles, Topock, Parker, Blythe, Yuma and permanently destroy the levees of the Imperial District, creating a channel into the Salton Sea which would probably be so deep that it would be impractical to reestablish the Colorado River in its normal course."

So much for the dam. What about the canal? Arizona was as interested as California. An ambitious plan was promulgated for a high-line canal that would divert a large portion of the river to irrigate millions of acres in Arizona. Engineers reported that this canal was not feasible from either an economic or an engineering standpoint. Next they investigated the proposed All-American Canal.

Imperial Valley was still receiving all its water for both irrigation and domestic purposes from the Colorado River by means of the Imperial Canal. Water was diverted at the Rockwood Heading, one mile north of the international boundary, and carried through Mexican territory to avoid the high mesa and sand dunes in American territory. The All-American Canal was to run entirely through this forbidding area. Water was to be diverted at Laguna Dam, 23 miles above, and carried for 75 miles straight across the desert into the valley. For 10 miles the canal would have to be cut through soft, shifting sand dunes 150 feet high. It would have to be made of concrete and be 50 feet deep. And it would cost at least $31,000,000.

Such were the results of the preliminary surveys submitted by Chief Engineer Weymouth to Director A. P. Davis of the Reclamation Bureau, and by him to Secretary

Fall of the Department of the Interior. Approved, the Fall-Davis report, as it was known, had then been sent to the Senate with recommendations for adoption.

By 1927, then, after eight years' work, the issues were clear on one of the most baffling problems facing the country. Here was the wildest river in America draining nearly one-twelfth of the whole United States, but flowing unchecked and for the most part unused into the sea. It threatened not only perpetually to inundate the rich Imperial Valley, but possibly by the subsidence of silt at the mouth to cause one of the greatest disasters in world history.

A Colorado River Project stupendous in magnitude was needed at once to avert certain disaster. If accomplished it would provide untold benefits for years to come. It would prevent the erosion of great areas of reclaimable land and stop the deposition of silt at the mouth of the river; provide electric power and water for both irrigation of land and domestic use by rapidly growing cities; and conserve and safeguard American interests by putting an end to the expedient of carrying water through Mexico and by ensuring a steady supply at all times. But such a project would cost millions. How should it be undertaken? Who was to do it? How was the bill to be paid?

Swing and Kettner's bill of 1919 had failed to be enacted. Hiram Johnson's bill of 1922 had failed. The Colorado River Compact of 1922, requiring the ratification of all seven states, had been declined by Arizona and could not be put into effect. Then Swing and Johnson had got together and presented three successive joint Swing-Johnson bills that failed to come to a vote.

Now Swing and Johnson worked out a new plan combining all the answers to the immense problem. Known as the Swing-Johnson bill, their fourth joint bill authorized the enactment of the Boulder Cañon Project Act. By its provisions the federal government would construct, operate and maintain a dam at Black or Boulder Cañon, a power plant near the dam, and the All-American Canal. Before any money could be appropriated or any construction work done, the Secretary of the Interior was required to make provision for revenues, by contract or otherwise, which would be adequate to pay for the entire project within fifty years. Also the Colorado River Compact was to become effective with the ratification of six states.

On December 5, 1927, the bill was introduced in both houses of Congress. Immediately it touched off a new and violent controversy. The key to it was the tremendous hydroelectric power to be generated at the dam.

Years before, when the Colorado River was being diverted for the first time into Imperial Valley, a young emigrant from Yugoslavia had come to Colorado. My own grandfather had known him well and had constructed for him an experimental shop in which to study the violent electrical storms and lightning discharges for which the region was famous. Nikola Tesla happened to be one of the world's great geniuses. Father of the alternating-current power system, of multiphase current transmission of high voltage, he ushered in the modern electrical power era of today. Through his discoveries Boulder Dam was to be built and paid for, and the Colorado River finally controlled.

For, in brief, the contracts for revenue sufficient to pay for the project were to be made for the sale of power generated at the dam. Control would thus be vested in the government. Wherefore the organized private hydroelectric power interests of the country set up a clamor. They would not stand for the government monopoly of its firm power, sold as falling water measured in voltage delivered at the switchboard, even to guarantee the financial integrity of the project. Congress was deluged by telegrams protesting against this feature of the bill. Such protests bore fruit in many devious ways.

Arizona claimed revenue from any development of the river, in the form of either taxation on the works or a royalty on the hydroelectric power generated. Her claim was based on the contention that the Colorado was navigable and that she was owner of the bed of the stream. Utah for a time also withdrew from the plan, contending that she was in accord with Arizona because of the discovery of oil in the bed of the river. However, if the government would turn all the power privileges at the dam over to private power interests, she would return to the six-state plan of ratification she had formerly approved.

The bill finally came up to vote. In a tense, dramatic ten days, in the face of determined filibuster by private power interests, the political struggle came to a climax. On December 14, 1928, the Boulder Cañon Act passed the Senate and a week later it was signed by President Coolidge.

The political fight had been won. Now the financial struggle began.

The Colorado River Board submitted a revised estimate of the cost of the projeit.

The dam in Boulder Cañon and a reservoir of 26,000,000 acre-foot capacity would cost $70,600,000; a 1,000,000-horsepower hydroelectric plant, $38,200,000; the All-American Canal, $38,500,000; and interest on these during construction, $17,700,000—a total cost of $165,000,000.

But before work on the project could be started it was necessary that the contracts be made for the sale of power insuring the payment of the entire cost within fifty years. They were not difficult to obtain: 27 applicants requested 322 per cent of all the power that would be available. The City of Los Angeles and the Southern California Edison Company each asked for the entire power output, and the Metropolitan Water District requested about one-half the total energy.

COLORADO RIVER PROJECT 327

The secretary of the interior was accordingly caught between the two horns of a new dilemma. The dam would bridge Arizona and Nevada, which were sparsely populated and desired power to build up their cities, but neither state was in a position to make a firm contract for the use of the power. On the other hand, the California companies offered the only definite means of financing the project.

After a formal hearing allocation of the total firm power was made as follows: the Metropolitan Water District, 36 per cent; the City of Los Angeles, 19 per cent; the Southern California Edison Company, 9 per cent; and Arizona and Nevada, each 18 per cent.

Contracts were entered into with the first three parties. But no contracts were executed on behalf of Arizona and Nevada, these states being given merely the right to exercise their options within fifty years.

Obviously, according to some proponents, California communities and private companies were to receive all the power; according to others, she was obligated to pay the entire cost of the dam.

Behind these contentions lay a new tangle and still another major addition to the huge project. The Metropolitan Water District had been organized as a public corporation the same week that the Boulder Cañon Project Act had been passed. It was composed of Los Angeles and twelve smaller cities within its metropolitan area. Supporting about 45 per cent of the entire population of California, they had joined together for the development of a water supply. Their concern was readily understandable. It took one acre-foot of water to support five urban citizens a year. The only adequate source remaining was the Colorado River 250 miles away. On it now depended the future growth of all these cities. The district therefore had proposed building another dam at Parker, 150 miles below Boulder Dam, and from its reservoir constructing a huge aqueduct to carry water to the metropolitan area. Construction was to be handled by the federal government, but the major part of the $220-million-dollar cost was to be borne by the Metropolitan Water District. The condition was made, however, that the United States agree to deliver from Boulder Dam enough water to fill its needs—approximately one million acre-feet per year.

Arizona literally hit the ceiling of her own blue sky. California was not only grabbing all the hydroelectric power, but all the water of the river in order to build up her own cities and orange groves while Arizona remained an unpopulated, barren desert. Arizona still claimed the Colorado River as her own by right of its being a navigable stream. To prove it the governor chartered the little steamboat *Nellie*, loaded her with the State Militia and steamed her upriver to Parker. Here for months the "Arizona Navy" remained obdurately anchored until the "City of Angels" agreed to cut Arizona in on her small share of water.

Uncle Sam was forced to step in. The Colorado River Compact had become effective with the six-state ratification under the Boulder Cañon Project Act, leaving Arizona the only standout. Yet Arizona's claim proved valid in court: California was demanding nearly 5,500,000 of the 7,500,000 acre-feet annually apportioned to the Lower Basin states.

By legislative action California now agreed to limit her use to 4,400,000 acre-feet, plus one-half of the excess and surplus. This was in turn divided by the Seven-Party Water Agreement of 1930 among Imperial, Coachella and Palo Verde valleys, Los Angeles and San Diego, and the Metropolitan Water District.

The long struggle with all its side-show squabbles was over. Contracts were ready to be let for actual construction on the various works comprising the Colorado River Project and constituting a total cost of over $400,000,000. Man at last was ready to subdue the wildest river in America.

CHAPTER 3

Boulder Dam

BOULDER DAM is the Great Pyramid of the American Desert, the Ninth Symphony of our day, and the key to the future of the whole Colorado River basin.

No other single piece of man's handiwork in this vast wilderness hinterland has epitomized so well during its construction all the strange and complex ramifications of our American Way—all its democratic faults and virtues, the political interlocking of local, state and federal governments, the meshed and rival economies of public and private enterprise, the conflicting needs of urban, agrarian and industrial groups. Finished, it stands in its desert gorge like a fabulous, unearthly dream. A visual symphony written in steel and concrete—the terms of our mathematical and machine-age culture—it is inexpressibly beautiful of line and texture, magnificently original, strong, simple and majestic as the greatest works of art of all time and all peoples, and as eloquently expressive of our own as anything ever achieved. Yet wholly utilitarian and built to endure, it is the greatest single work yet undertaken to control a natural resource dominating an area of nearly a quarter million square miles. . . .

Boulder Dam, the biggest dam on the face of the earth

and the first major work in the Colorado River Project. Already the blue chips were down. The United States Reclamation Service had finished the specifications. Government lawyers had condemned 150,000 acres above the site. The secretary of the interior had planned the route of the little construction railway running from Las Vegas, Nevada, and on July 7, 1930, the traditional silver spike had been driven.

It was now nearly eight months later, and in two days bids for the construction of the dam were due. Yet no one knew if there was a man or company in the country big enough to ask for the job.

No one but a small group of men in the St. Francis Hospital, San Francisco. They were gathered about the deathbed of an old man of seventy-two. A muffled knock on the door broke the silence. Into the room came a younger man of forty-eight wheeling a strange contraption. The eyes of the old man lighted up as it was placed beside his bed. It was a scale model of Boulder Dam. The younger man had made it just as he was to construct it full size later. He knew it. All the others felt it in his voice as carefully now he went over its every detail. At last he finished. For a moment there was silence. The old man on his deathbed waved his hand in approval.

"One last thing," another spoke up. "What'll we add to the cost for profit?"

"Profit? Twenty-five per cent! Tidy up the estimates and get the bid in."

"Right!" Quietly the men shook hands and left. W. H. Wattis sank back content into his pillows. He had seen the crowning achievement of his life work as clearly as if it already stood across the turbulent Colorado.

Behind it lay a story almost as fabulous as the dam itself—his own and that of his companions who were to build it.

Eighty years before, an Englishman had joined the California gold rush. He got no farther than Utah. His two sons, W. H. Wattis and E. O. Wattis, became Mormons and contractors, helping to build the old Colorado Midland Railroad that ran through the mountains from Colorado Springs to Ogden. Here at its western terminus the two Wattis brothers founded the Utah Construction Company which built the Hetch Hetchy dam that impounded San Francisco's water supply. E. O. Wattis was now approaching the age of seventy-six; W. H. Wattis, seventy-two and ailing; and they had made as president of the company their 60-year-old cousin, Lester S. Corey. But Boulder Dam was due and they determined to build it.

The project was too big for the Utah Construction Company to handle alone. They decided to appeal to the Morrison-Knudsen Company, Inc., of Boise, Idaho. This company had been formed in 1912 by Harry Morrison and Morris Knudsen, who in 1925 had taken into the firm Frank T. Crowe. Crowe had spent most of his forty-eight years in river bottoms. With Morrison-Knudsen he had just finished building the Guernsey dam in Wyoming and the Deadwood dam in Idaho. Previous to this he had been general superintendent of the United States Bureau of Reclamation and in 1919 had made one of its first rough estimates of Boulder Dam.

The two firms got together. Boulder Dam, they figured, would cost from 40 to 50 million dollars. At least $5,000,000 working capital would be needed to start it. The two old Wattis brothers offered to put up $1,000,000 for the Utah Construction Company. Morrison-Knudsen agreed to chip in $500,000. Together they approached a third company.

The J. F. Shea Company, Inc., of Los Angeles had been founded in 1914 by a plumber and his son, Charles A. Shea. The son soon became one of the best sewer and tunnel experts on the Pacific coast, and the firm secured contracts for laying the water-supply lines for San Francisco, Oakland and

Berkeley. The Shea Company now jumped at the chance to get in on Boulder, agreed to ante $500,000, and suggested that the Pacific Bridge Company of Portland be called in.

This fourth company was famous for its underwater work. It had driven the piers for the first bridge across the Willamette River at Portland, and working with Shea had laid the water line across the Mokelumne River. Heading the company was W. Gorrill Swigert, an Oregonian. He too agreed to put up $500,000.

A total of $2,500,000 had now been secured—just half of the $5,000,000 required, when W. H. Wattis took to bed. It was found he had developed a cancer of the hip. Nevertheless, he decided to stay by his proposal and urged his associates to find another company to finance the remainder.

The fifth company picked was MacDonald and Kahn, Inc., of San Francisco. No stranger pair of partners could have been found. One of them, Felix Kahn, was a quiet, shrewd, 61-year-old Jew born in Detroit, the son of a rabbi. The other, Alan MacDonald, was a fiery and impetuous Scot who had been fired from fifteen consecutive jobs before teaming up with Kahn. From the start they got along, building the Mark Hopkins Hotel of San Francisco and some of the largest office buildings on the West Coast, totaling some $75,000,000 in construction. They were already interested in Boulder Dam and agreed to add another million dollars to the pool.

It was still $1,500,000 short; time was slipping by. Meanwhile another group of men had become just as interested in the biggest dam in the world.

Two of these were W. A. Bechtel, founder of the construction firm W. A. Bechtel Company of San Francisco in 1900, and his son, S. D. Bechtel.

Another was Henry J. Kaiser of Oakland, 61-year-old head of the firm bearing his name. Years before, Kaiser as a young man from Upper New York State had gone to California hunting construction work. "Dad" Bechtel, "a tall beefy man with a bull-like roar" took him in and pushed him up the ladder to success. Kaiser was a rapid climber. He soon established his own company, and became national president of the Association of General Contractors. At the present time he was working on a subcontract secured from Warren Brothers, a construction company of Cambridge, Massachusetts, for building a 20-million-dollar highway in Cuba.

In 1930 Kaiser was in Cuba finishing the job when Bechtel arrived to propose Boulder Dam as their next joint project. Kaiser agreed, but suggested they talk it over with John Dearborn, chairman of the board of Warren Brothers.

As representatives of their three companies they all met to discuss their own plans, and to consider those of the Wattis-Morrison group about which they had learned. Obviously the two groups could not be rivals; the dam was bigger than both of them. Accordingly the Bechtel, Kaiser and Warren firms teamed up with the preceding five firms as the sixth company, adding the remaining $1,500,000 split between them.

For the new combine Kahn suggested the appropriate name of Six Companies, Inc., called after the famous tribunal to which the Chinese tongs in San Francisco had submitted their differences in preference to warring with hatchet men.

In February, 1931, it was so incorporated in Delaware. W. H. Wattis on his deathbed in San Francisco was selected president; Dad Bechtel and E. O. Wattis, first and second vice-presidents; Shea, secretary; and Kahn, treasurer. Crowe worked out all the details and made his own scale model from his blueprints.

Approved in Wattis's hospital room, the last preliminaries were then agreed upon. Next night in the Cosmopolitan Hotel, Denver, Crowe made up the final bid of $48,890,000 from three separate estimates. The following day, March 4, 1931, it was submitted.

The weeks dragged by until a decision was rendered. Few of the men waiting so patiently could guess even then how it was to shape their entire future. Four of them were to die without seeing the dam finished: W. H. Wattis, in his hospital bed six months after it was bid; E. O. Wattis, soon after; Dad Bechtel, in 1933 while on a trip to examine Russia's subway system; and Alan MacDonald, in 1935.

The rest of the group was to stay together as the Six Companies and gain world renown. After some quarreling, Shea was placed in charge of field construction; Kahn in charge of finances, legal affairs and housing; the younger Bechtel of purchasing, administration and transportation; and Kaiser, having a knack for making them co-operate, was made chairman of the board. Finishing Boulder Dam with a profit of $10,400,000, they sank the piers for the Golden Gate Bridge; worked on the Bay Bridge; built Bonneville Dam and finished the Grand Coulee Dam on the Columbia River. Today they are finishing a length of the Alaska Military Highway; have built underground storage tanks for Pacific Navy air bases; have helped to raise warships sunk at Pearl Harbor.

Frank Crowe, who as superintendent of construction did more than any other single man to build Boulder Dam, made his fame and fortune of nearly $300,000 from his modest salary and 2.5 per cent of the gross profit.

Henry J. Kaiser from Boulder, the world's highest dam, climbed swiftly to world eminence. Almost "psychopathetically" power-mad and bluff as his old boss, he first built Permanente, the world's biggest cement plant, to break the cement combine. Foreseeing the war crisis he next built with $106,000,000 of RFC funds a new steel mill at Fontana, near Los Angeles, establishing himself in the steel industry. Finally with the war and his own knack for obtaining publicity, he rose to his present fame as the world's fastest shipbuilder. On his first 120-million-dollar order for 60 ships, he built the first one in 196 days. This time was then cut to 25 days, and Kaiser finally was launching a new Liberty ship every 10 hours. The great postwar reconstruction period offers still greater opportunities to him and his companions of the Six Companies.

But as yet they were all still chewing their nails over the outcome of their bid for their first big job. It was not long in coming. There had been only two other bids on Boulder Dam. One was $5,000,000 and the other $10,000,000 higher; the Six Companies' bid was taken. Wattis and Morrison had hit the nail on the head in figuring the minimum working capital that would be required. For the surety companies, which first demanded $8,000,000 for underwriting the job, now agreed to accept the $5,000,000. And on April 20, 1931, the Six Companies received notice to begin. Work with dollars and decimal points was over; it was time to pick up shovels.

An old engineer once told me that half of any construction job was decidin' to do it, and the other half was figurin' how to do it right; then all you had to do was git it done. So it is. For a song may be resung; a book rewritten; a new medicine given as an antidote for the old. But a bridge or a dam stands or falls on its builder's preparation. Nothing bears this out more than Boulder Dam. A single fault would be fatal. And after twelve years or more of careful thinking and figuring, the work progressed at record speed. The full story of its construction lies only in its blue prints. For us there was but a series of flashing pictures before it was done.

Swirling westward out of Grand Cañon and gorged with its cutting silt, the Colorado makes a sudden turn due southward to form the boundary between Nevada and Arizona. Its channel is a series of deep, steep-walled cañons. Not far below the turn is Black Cañon, dark and forbidding as its name. The stark black and purple rock cliffs on each side rise a quarter of a mile high from the water's edge. Beyond these the tawny desert stretches away unbroken save for a rough dirt road straggling toward the west. It terminates near the upper end of the cañon at a break in the rock wall. Here a tent is pitched. In it is camped an old couple making coffee for some booted visitors whose motorboat is drawn up on the sandy bank near by.

This was Black Cañon when I saw it early in 1930. We got into the boat and pushed off. Instantly the current grabbed it and sucked us deep into the cañon. On the westward wall appeared some white bench marks one of the engineers had painted the week before.

"How far above the waterline do you figure the lowest one is?" he asked as we rushed past.

"Five feet!"

"Twenty-five!"

It was the 100-foot mark, and it designated the site of Boulder Dam. Immediately we turned about and began to fight our way back upriver. The trip down from the landing had taken about ten minutes. The return trip took almost a half hour.

Nothing, certainly no statistics, could portray as did this short, casual ride the entire problem of building Boulder Dam: bridging sheer, unbroken cañon walls so high that they distorted perspective; the lack of even a sand bar for initial footage; the desolate desert on each side without housing or transportation facilities; and greatest of all, the terrific current of the silt-choked river.

That evening we chugged by Ford back over the desert road to Las Vegas and lay sprawled out on the grass drinking cold beer. Across the dusty road stood a line of cribs. Outside each, in tipped-back chairs, lazed the girls waiting for chance customers and hulloing at passers-by. The sun set.

The lights came on. A few cars rattled into town with ranchers and their families. One of my friends stood up. "My wife—I promised to take her to the movies." We walked downstreet with him. A block of bright-lit bars and gambling casinos was filling up with tourists. Riproaring, gambling hell-for-leather, non-Prohibition style, as they said —all against national law. This was Las Vegas before Boulder. The last frontier town in America, but really dying in its boots.

A year later I was sent back to make a quick survey of communication facilities. There was a little local switchboard manned by three telephone operators drinking Coca-Cola who knew all voices and required no numbers to make connections. Rural lines were strung along barbed-wire fences. Long-distance calls were rare as Christmas. . . . Necessary improvements, to put it mildly, were indicated. And quickly. The Boulder contract had been let. Six Companies' engineers were trying to call Los Angeles, New York.

Suddenly Las Vegas woke up. Overnight she changed completely. One could not record all the changes; one could only stand gaping at the tornado of activity.

Men and material poured into town. To house Six Companies' workmen a new model city was built close to the river. It bore no resemblance to the rude construction camps in the Wyoming oil fields I had lived in not ten years before. Since then something new had entered the American scene on the heels of the great depression—social consciousness. Not everyone was aware of it, or else they persisted in maintaining a squalid Rag Town of scrap lumber not far away merely to uphold the tradition of all mining, oil and construction projects throughout the West. But Boulder City had paved streets, stores, tidy bungalows and air-conditioned barracks, enormous mess halls, garages, machine and service shops. Telephone and telegraph service were provided. A paved highway and a railroad were built in from Las Vegas, and an electric transmission line from California to furnish light and power.

Supplying water was the chief problem. Out in the desert lay only one oasis where years ago my cousin's father-in-law had discovered an artesian spring. Homesteading a section around it, he had patiently broken and irrigated land, built a rambling ranchhouse. A small orchard was his chief pride—rows of apple, cherry, plum and pear trees. Lately he had constructed a tiny greenhouse. One evening a group of engineers called upon him, asked to see the spring. There it was: a steady bubbling column of clear fresh water.

"How much? We want it. . . . Name your price."

The old man looked across his small green island to the pelagic sea of sand around it. "I reckon you'd have to buy my house and barn and greenhouse, my fruit trees, and maybe my forty years of work and peace of mind to boot," he said slowly. "Without this spring there wouldn't be any of them left."

So, instead, water was supplied Boulder City from the river by means of a $500,000 water supply plant that pumped, purified and distributed 50,000 gallons daily at a cost of one-half cent a gallon.

Meanwhile "5,000 men in a 4,000 foot cañon" were at work on the dam. First the sheer rock walls had to be cleared over 2,000 feet high of loose rock and boulders. Work began at the top as "cherry pickers" carrying jack-hammer drills and dynamite were let down with ropes. This was one of the most dangerous tasks of the entire job. A miscalculation of distance and a man would drop a quarter mile into the river; of time, and he would be blown to bits by his own charge of explosive. Above these was the terrific desert heat reflected from the narrow cañon walls, which dehydrated the hard-laboring workers quickly. They were soon required to carry water bags at all times—and to empty them regularly.

BOULDER DAM 339

Owing to all such precautionary measures taken, only 110 men were killed during the entire construction.

Once the walls were cleaned and safe to work under, a spectacular job began. The bottom of the river bed had to be bared as well, and kept clean and dry during the whole period of construction—and bedrock lay under the rushing red river and 200 feet of sand. To divert the river, two by-pass tunnels were cut through the cañon walls on each side. Each of the four was a mile long, 50 feet in diameter and lined with 3 feet of concrete. These alone required 1,200 men one year to build, riding air-drill carriages into the tunnels and blasting their way with dynamite.

Work progressed at tremendous speed, day and night, without cessation. At the end of the first year Six Companies was in the clear. The men were ahead of schedule, had got back all their working capital and made a million dollars in profits.

On November 13, 1932, the by-pass tunnels were done. The Colorado began flowing through the cliffs, forsaking the channel it had been cutting for perhaps 15,000,000 years.

One midnight, soon after, I stood on the bed of the river. The vast chasm seemed a slit through earth and time alike. The rank smell of Mesozoic ooze and primeval muck filled the air. Thousands of pale lights, like newly lit stars, shone on the heights of the cliffs. Down below grunted and growled prehistoric monsters—great brute dinosaurs with massive bellies, with long necks like the brontosaurus, and with armored hides thick as those of the stegosaurus. They were steam shovels and cranes feeding on the muck, a ton at a gulp. In a steady file other monsters rumbled down, stopping just long enough to shift gears while their bodies were filled with a single avalanche, then racing backward without turning around. From the walls above shot beams of searchlights, playing over this vast subterranean arena. They revealed puny pygmies scurrying like ants from wall to wall; mahouts, naked to the waist, riding the heads of their mounts, standing with one foot on the running board and peering over the tops of the cabs while driving with one hand. And all this incessant, monstrous activity took place in silence, in jungle heat, and as if in the crepuscular darkness of a world taking shape before the dawn of man. . . .

Meanwhile the largest concrete-mixing plant in the world had been built high on the Nevada wall. Its daily output was equivalent to a stream of concrete 20 feet wide, nearly 1 foot deep and 1 mile long, and it was to flow steadily for 2½ years. On June 6, 1933, it began flowing: replacing nearly 6,500,000 cubic yards of excavated rock and sand with 3,250,000 cubic yards of concrete—more than the United States Bureau of Reclamation had used in the past quarter of a century. Into it dipped the first steel bucket, scooping up 16 tons. Swiftly it was carried across the gorge on cables and then let down to dump its contents in the frames. Pouring had begun.

At first the structure looked little like a dam; more like a vast jumble of wooden boxes filled with concrete. There was reason for such design. Had the dam been built solid, 125 years would have been required for it to cool by itself, and under the great stresses of expansion it would have cracked and split apart. Hence to cool it an ammonia refrigerator plant was built. From it 662 miles of tubing were run between and through these immense blocks, carrying water just above freezing to maintain the rate of cooling at the right temperature.

Swiftly these water-cooled boxes heaped higher across the gorge. The wooden frames were removed. There it stood: a huge, rough, pyramidal block of concrete 660 feet thick at the bottom and 45 feet thick at the top, 1,282 feet long at its crest between the cañon walls, and 727 feet high— over half again as high as Washington Monument. But still honeycombed with narrow passages, galleries and ducts. One walked through the base of the dam, the length of three city blocks, or rode a small elevator to its crest, the height of a 60-story skyscraper. A labyrinth of cold passages; the dark interior of a great pyramid; the heart of a stone mountain; the depths of a mine—it was all of these. Then the tubing was removed, the corridors were pumped full of concrete, and the dam was sealed to resist the 45,000-pound-per-square-foot water pressure on its base.

Inside both the Nevada and Arizona walls of the cañon the diversion works were now finished. On each side, upriver, were two great intake towers 395 feet high. From the bottom of these, 30-foot steel penstocks ran through the cliffs, branching laterally on the downriver side to carry water to the powerhouse or to release it through the outlet works farther down. Every dam must also have a spillway to operate as a safety valve in preventing the reservoir from overflowing the dam in case of a flood. Boulder had two, one each on the Nevada and Arizona sides, either of which could dock the biggest ship afloat. From each of these was bored through the cañon walls a spillway tunnel 50 feet in diameter and 2,200 feet long, emptying into the river channel below the dam.

Finally at the bottom of the dam was built the world's largest powerhouse to date. U-shaped, each of its two wings was a city block long and high as a 20-story building. In it were to be installed two generators of 40,000 kilovolt-ampere capacity—as large as any in the world—and fifteen of 82,500 kilovolt-ampere capacity, run by two small turbines of 55,000 horsepower and fifteen large turbines of 115,000 horsepower each. From here across the desert to Los Angeles, 250 miles away, began to march the great steel Martian towers carrying high-voltage transmission lines to supply 3,000,000 people with power.

In May, 1935, one year ahead of schedule, Boulder Dam

was completed. An unequaled engineering achievement, it had been constructed in exactly 4 years and 354 days.

During the whole period there had been but one hitch. The "Curmudgeon of the New Deal," Secretary Ickes, charged the Six Companies with 70,000 violations of the eight-hour workday law and authorized a fine of $350,000. Kaiser rushed to Washington. With his knack of using publicity and pulling the proper political strings, he hired a publicity man, drew up a pamphlet *So Boulder Dam Was Built* to mail to members of Congress, and spoke over the radio to the world at large. As a result of his campaign the fine was reduced to $100,000.

The by-pass tunnels were now plugged, and the Colorado began backing up against the dam. In March, 1936, the government took over Boulder, and on September 11th President Franklin D. Roosevelt pushed a golden key starting the first generator of the initial installation of four. It was not too soon. Hitler was preparing to march through Europe, beginning World War II. The United States was to become involved, and the West Coast to need every ounce of power available for building bombers, tanks, guns and ships. Hence Boulder's ultimate capacity of 1,835,000 horsepower, originally planned to be reached in forty years, was to be attained in five in order to meet the national emergency.

But within a year, for 565 miles between the lower end of Grand Cañon and the upper end of the Gulf of California, the face of the land had been forever changed.

Below the dam lay 2,000,000 acres of desert, drier than Sahara, which would be reclaimed.

Above the dam was swiftly forming the world's largest reservoir, Lake Mead, named after the commissioner of reclamation. There was room for it to grow, as he had retained for it all land 115 miles upriver below 1,200 feet elevation. The lake would need it. Capable of impounding for two years the entire flow of the river averaging 15,000,000 acre-feet annually, this reservoir would cover the entire state of Connecticut with water 10 feet deep.

A strange lake, indeed, in the middle of the desert. Open to navigation for 115 miles, it will reveal some of America's most magnificent cañons, till now known only to a handful of daring travelers—Black Cañon, Travertine Cañon, Iceberg Cañon and the lower reaches of Grand Cañon. Already it is being used by a few old desert rats who boat their ore from isolated mines to the rail connections at Boulder, instead of packing it in by burros. The lake is being stocked with fish. Lying on one of the great migration lanes between Canada and the Gulf of Mexico, it is becoming a refuge for thousands of flocks of wildfowl. And it is being made into a vast recreational area for the half million tourists who visit it every year. The lake may well become a national park.

Six hundred feet below its surface lie the ruins of Lost City—Pueblo Grande de Nevada, a prehistoric pueblo believed to be over a thousand years old. Submerged too is the little Mormon village of St. Thomas. They are symbols that can be interpreted two ways: as graves of men's fears of the great red river or as testimonials of its power to obliterate ultimately all the handiwork of man.

CHAPTER 4

Canal, Aqueduct and Treaty

The All-American Canal

MEANWHILE construction had begun on the second group of operating works. As a tasty reminder of what they were to accomplish, the first cantaloupes of the season to be shipped from Imperial Valley were sent to President Roosevelt by air mail. It was March 23, 1934—a new record. The new All-American Canal was to ensure many more.

Four miles above Laguna Dam, near Yuma at the junction of the Gila, a new dam was built on the Colorado. Imperial Dam was an Indian diversion weir 25 feet high and 1,770 feet wide at its crest. With a flood capacity of 259,000 second-feet, it will eventually raise the water 20 feet above river level and create a backwater basin for 15 miles upstream.

Connected with it on the west end were installed desilting works, consisting of six parallel settling basins, each 110 feet wide at the bottom and 1,200 feet long. Above the intake gates were huge trash racks which removed the debris floated down by the river. Then the silt was settled and sluiced back into the river below the dam. Any five of these basins could be used while the sixth was being sluiced. All the gates were electrically controlled and could be operated from a central switchboard.

From these began the All-American Canal—a concrete-lined ditch receiving 15,000 second-feet of debris-cleared and desilted river water. For several miles it ran southward, parallel to the river, past Pilot Knob where water had been diverted into the old Alamo Canal at the Rockwood Heading. Then it turned westward and ran roughly parallel to the Mexican border across the sand hills and the mesa, dipping down into Imperial Valley.

The total length of the main canal was 80 miles. Some 130 feet wide at the bottom and 196 feet wide at the water level, it required a total excavation of 53 million cubic yards. On each side where it ran through the sand dunes a berm, or shelf, 15 feet wide was excavated to prevent sand from falling into the canal. To control the sand-blow the dunes were sprayed with oil, and on the high shifting crests a growth of vegetation was started.

Across the flat, baking desert, with the dry chocolate hills shimmering on the horizon. Past the old Mexican border town of Algodones, once a screaming lane of cantinas and casinos, and now almost completely gone. Across patches of the old plank road. Through the rolling, sheer-white sand dunes, America's little Sahara, where remnants of *Beau Geste*'s storied fort still stands and every romantic movie of North Africa was made. It is an unbelievable canal, a concrete-lined river of bright blue water 16 feet deep rushing through heat and sand at the rate of nearly 4 feet per second. The Hohokam of Arizona, those ancient Canal Builders, would have believed it impossible. Even for us it was a modern miracle. And another longer branch is yet to be completed. Tapping the main canal midway across the desert, it will extend northwest up past Salton Sink for 135 miles into Coachella Valley.

At five points along the course of the canal were sufficient drops in elevation to provide development of hydroelectric power. The first of these was at Pilot Knob, where the wasteway between the canal and the old Rockwood Heading drops 55 feet. The other four lay between the west side of the sand hills and the valley. Together they provided power drops totaling 175 feet. With power plants at all points, the full capacity will total 80,000 horsepower, with an output of 4,000,000 kilowatt-hours per year.

Built like Boulder Dam under terms of the Swing-Johnson bill, the total project cost slightly less than the original estimate of $38,500,000. This money was appropriated by Congress and advanced by the federal treasury without interest. It was to be repaid over a period of 40 years by the lands to be benefited: 1 per cent annually for the first 5 years, 2 per cent for the next 10 years, and 3 per cent for the remaining 25 years.

The actual cost to the farmers would thus be about 30 cents an acre annually for the first 5 years, 60 cents for the next 10 years, and 90 cents for the remaining 25 years.

The sale of power will reduce these payments as this income is to be applied toward payment of the loan. The desilting works also will save Imperial Valley up to $1,400,000 each year in the cost of dredging silt from ditches.

A cheap development indeed. Already the Imperial Irrigation District deserves its name as "America's Winter Garden." It is the largest irrigated area in North America, larger than the total farmed area of the state of Delaware, and it produces a major crop for every month of the year. Over 523,000 acres are now under cultivation and 838,000 more will be irrigable in the future. There are 2,900 miles of canals and laterals, and 10 towns below sea level with a total population of 65,000.

All lying within the heart of the Colorado Desert. All developed within a mere forty years. It is the "fantastic folly of an old man," the dream of a sick and broken physician that at last has come true.

The Colorado River Aqueduct

The third of the major works in the project was no less stupendous. It involved building another dam on the Colorado and from it constructing an aqueduct 242 miles across the desert to carry 1,605 cubic feet of water per second to 3,000,000 people in 13 cities within an area of 2,200 square miles.

Only the ancient Romans, fresh from the scene of their own great aqueducts, could have appreciated the magnitude of such a task. For us a single sight of the long serpentine pipe crawling across deserts, boring through mountains, is enough. Like the other miracles we can accept it calmly without the statistics of its construction. Its whole story is told when we turn on the tap. Nevertheless, if for some figures have a meaning, these are big enough.

As Boulder is the highest dam in the world, so Parker is the "deepest"—233 feet of silt and sand having been removed before its foundation rock was reached. It is located near Parker, Arizona, about 150 miles south of Boulder Dam. Its site is a narrow rock outlet on the Colorado just downstream from the mouth of the Bill Williams River.

Parker Dam is a concrete arch with five huge sluice gates, each 50 by 50 feet, on the crest. It serves three main purposes. It protects the land below from the flash floods of the Bill Williams; provides 135,000 horsepower of electric energy; and raising the water 72 feet, it provides 717,000 acre-feet of storage water to be pumped through the aqueduct as needed.

The Colorado River Aqueduct leads from it to the Cajalco Reservoir near Riverside, California, which provides terminal storage of 225,000 acre-feet. From this a distribution system spreads water to all the consuming centers.

The main aqueduct is 242 miles long. It winds across the upper end of the Colorado Desert, through desolate and parched mountains, and through San Gorgonio Pass which separates the Colorado and Mojave Deserts, the ancient pathway to the coastal plain. To carry supplies for its construction, 153 miles of surfaced highways and 250 miles of dirt roads had to be built; 477 miles of high-voltage power lines were needed to provide power, and 1,273 circuit miles of telephone lines. And to supply construction crews, camps and hospitals with water, 180 miles of water-supply lines with wells and pumps were required. During its construction 8,728 men were employed at one time—more than were required to build Boulder Dam.

The aqueduct itself was simple enough—a mere pipe line to carry water, a sizable job of plumbing, if you will. But it had many variations: lengths of concrete-lined canal, covered concrete conduit, siphons and tunnels. It had to bridge arroyos and gullies, snake through passes and cañons, crawl over hills, slide around some mountains and bore through others. There were 144 siphons, up to 12 feet in diameter, with a total length of 28 miles. Included in its 92 miles of tunnels were the San Jacinto, Whitewater, Eagle Mountain, Iron Mountain, Coachella and Cottonwood tunnels which had to be jack-hammered by half-naked, sweating men through the solid rock of gaunt, bare peaks. Finally, as the whole region was seamed with earthquake faults, the course of the entire aqueduct had to be plotted so that it would cross these faults at right angles and be flexible enough to avoid being broken by the frequent tremors.

To lay such a long, devious line across a level plateau would have been a tremendous job. This one was made still bigger by the necessity of lifting the entire flow as high as 441 feet at a time, and a total of 1,617 feet. Even this was made more difficult by the tremendous amount of water to be forced through it and lifted—1,110,000 acre-feet a year,

CANAL, AQUEDUCT AND TREATY 349

1,605 cubic feet a second. The pumping problem, in short, was the greatest ever undertaken.

Five pumping plants were installed along the course of the aqueduct: the Intake, Gene, Iron Mountain, Eagle Mountain and Hayfield lifts. Each plant contained an initial pump installation of three units. Selection of the equipment necessitated the construction of a complete pump-testing laboratory at the California Institute of Technology, and leading manufacturers co-operated in the design of test models for two years before selection.

Installation of great electric motors to run these huge pumps required enormous power—375,000 horsepower. The Metropolitan Water District, by virtue of its contract, was allocated 36 per cent of the firm energy available from the power plant at Boulder; first right to all secondary energy, and to any unused portion of the 36 per cent allotment to Arizona and Nevada. Hence new power lines were built down from Boulder Dam to the five pumping plants. Supported on rigid steel towers, the conductor wire, over an inch in diameter, was made of aluminum except for its 10-mile stretch across Danby Dry Lake. Here copper was used, as it offered greater resistance to corrosion from alkali dust. Its transmission potential was 230,000 volts, stepped down to 6,900 volts for use at the pump motors. Carrying an ultimate 280,000 kilowatts, the electricity was sufficient to light a city larger than Los Angeles.

One other task remained: a filtration and softening plant in which the hardness-forming minerals in solution in the water were removed by the use of lime and zeolite, and the water then filtered to render it fit for irrigation and drinking.

Work was started on Parker Dam in October, 1934, and completed in August, 1938. There had been but one interruption, due to the presence of the "Arizona Navy" pending settlement of California's legal controversy with California over the use of the river.

One year later, in the fall of 1939, delivery of water began to Los Angeles and twelve other cities in the Metropolitan Water District: an area comprising the richest four counties in value of crop production in the United States, 3,000,000 people, and 2,200 square miles of reclaimed desert. Without it, future growth seemed impossible.

It was altogether appropriate, then, that the region should have borne the cost of Parker Dam and the Colorado River Aqueduct. It did. Unlike Boulder Dam and the All-American Canal, this third major project was paid for by the thirteen cities within the region it served.

Parker Dam was constructed by the United States Bureau of Reclamation with funds furnished by the Metropolitan Water District of Southern California. The government retains title to the dam and operates it for the district, each retaining half the power privilege. The total cost of dam and aqueduct was $220,000,000. Payment was ensured by a bond issue, and funds to carry on the construction were supplied by loans from the Federal Reconstruction Finance Corporation and the Federal Emergency Administration of Public Works.

The physical work was finished on all projects. Perhaps never before in history had so much money been poured into a single area; nearly $425,000,000 within an area of less than 200 miles radius. But, strangely enough, one last argument over their use was already brewing.

United States-Mexico Water Treaty

The completion of these projects aggravated rather than relieved an age-old international headache caused by the Colorado. Today, as these lines are written, a final effort is being made to cure it. A treaty between Mexico and the United States, providing for the distribution of the waters

CANAL, AQUEDUCT AND TREATY 351

of the Colorado, is pending ratification by both countries. It is one of the most important of its kind in the world. Yet much more is at stake than a division of water. For it has set off a controversy that intrudes into the whole field of inter-American relations.

Behind it lies a century of juridical dispute and international co-operation that has few parallels.

In 1848 and 1853 the first boundary treaties between Mexico and the United States defined and fixed the boundaries between the two countries. These were formed in part by the courses of three rivers—the Rio Grande, Colorado, and Tijuana.

The Boundary Convention of 1884 then met to consider the problem of their shifting channels. It provided that these waterway portions of the boundary should forever remain as previously agreed—i.e., the boundary should follow the middle of the original channel bed, "unaffected by alterations caused by erosion and deposits of alluvium, or any change wrought by the force of the current whether by cutting a new bed or deepening of another channel." The later Convention of 1889 provided means of carrying out these principles. It established an International Boundary Commission, with members appointed by both countries, to render decisions on all questions and differences submitted to it.

By 1905 another problem had developed over the small tracts of land called "bancos" which were continually being separated from one country and attached to the other by changes in the course of the river. Accordingly a Banco Convention was called. It extended the commission's powers, authorizing it to survey these bancos and eliminate them from the effects of the Convention of 1884 by transfer of their sovereignty and jurisdiction.

Meanwhile still another question had been posed by the growing use of water from the Rio Grande for irrigation and domestic purposes. In 1924 the United States Congress approved the establishment of another commission to study this new problem: the International Water Commission. Mexico, however, refused to discuss the Rio Grande unless the Colorado also was considered. After eight years of discussion all negotiations failed. The powers and functions of the International Water Commission were transferred to the International Boundary Commission; and in 1935, by an act of Congress, this commission was authorized to co-operate with the representatives of Mexico in making a full study of the uses of all three streams—the Colorado, Rio Grande, and Tijuana—as a basis for the negotiation of a treaty between the two countries.

Mexico agreed. A new study was begun. As expected, the Colorado offered at once the chief difficulty for settlement.

Under terms of the Colorado River Compact of 1922, 7,500,000 acre-feet of water were apportioned annually each to the Upper Basin and the Lower Basin. In addition, the Lower Basin reserved the right to increase its use by 1,000,000 acre-feet yearly if needed. It was further agreed that if any water was allocated to Mexico by treaty, it was to be supplied from the surplus above this total of 16,000,000 acre-feet; and if this was insufficient, the deficiency was to be borne equally by both basins. As the total flow averaged 17,400,000 acre-feet, this left but 1,400,000 to be allocated by the present treaty.

At the time of the compact, these provisions seemed entirely fair. About 80 per cent of the Colorado's flow originated in the high mountains of Colorado and Wyoming. Included within its entire drainage basin were 244,000 square miles in the United States compared to only 2,000 square miles in Mexico. Moreover, most of the water was wasting unused through Mexico into the Gulf.

But by 1943 the picture had changed considerably. The various works of the Colorado River Project were completed.

CANAL, AQUEDUCT AND TREATY 353

Immense amounts of water were diverted through the All-American Canal and the Los Angeles Aqueduct. Boulder Dam was backing up the river. And 1,312,000 acres were under irrigation in the Upper Basin states, and 1,323,000 acres in those of the Lower Basin.

In Mexico similar changes were under way. It had been proposed to open up to development both the Mexicali Valley and the Mexican Delta of the Colorado River just as the United States had opened Imperial Valley years before. In 1937 the National Irrigation Commission had taken charge, and in 1943 had begun to build a network of canals. Some 300,000 acres were already under irrigation, and 1,000,000 acres were readily irrigable—almost as much as in each the Upper and Lower basins above the border. But now the Mexican farmers woke up to what was happening; Boulder Dam was impounding all the water. In April, 1943, they appealed to their president, pointing out their impending agricultural ruin and asking him to take the necessary steps to save the whole irrigation program from catastrophe.

Mexico therefore demanded 3,600,000 acre-feet of water from the Colorado annually. The United States offered 750,000—the maximum amount that Mexico had been able to use to date.

A deadlock was reached. Clearly there was not enough water for all new projects for expansion of irrigated land in both countries. For during that year, 1943, only 1,800,000 acre-feet of water were diverted to Mexico from the United States. After another eight years of discussion the International Boundary Commission had failed to make a decision. Negotiations between the countries broke down.

It was obvious that a major treaty was necessary, and the United States Department of State undertook to draw it.

There were many precedents for evolving such a treaty. The treaty of 1599 between Venice and Mantua relative to the utilization of the Tartaro River provided for the appointment of a commission. The Central Commission of the Rhine was set up in 1815, and the European Commission of the Lower Danube in 1856. Treaties had been signed by France and Italy relative to the River Roya; by Spain and Portugal regarding the River Duro. As late as 1929 a treaty between Egypt and Great Britain was drawn, governing the use of the Nile for irrigation.

All these treaties were limited in scope. They designated commissions, fixed boundaries, provided for irrigational use of the water. But they preceded or overlooked the possibilities of enormous power development and did not make provision for great international works of permanent character.

The proposed United States-Mexico water treaty would have to include all these factors. It would have to evolve principles for international co-operation in dividing the water and provide the administrative machinery for development of the river's power resources. As such it might become a model for future treaties governing such great international streams as the Amazon, La Plata and Congo as these in turn were developed for the use of mankind.

The State Department accordingly began conferences with what became known simply as the Committees of Fourteen and Sixteen representing both water and power interests of the Colorado River basin, two members being chosen from each of the seven states concerned and two from the contractors for power derived from Boulder Dam. Out of these meetings evolved a treaty meeting all requirements and approved by the majority vote of all representatives. The State Department then reopened negotiations with Mexico, and on February 3, 1944, the treaty was signed by both countries.

Briefly, the treaty guaranteed Mexico 1,500,000 acre-feet of water annually, plus an additional 200,000 acre-feet in surplus years. This amount, it will be remembered, is approximately the difference between the average total flow of 17,400,000 acre-feet and the 16,000,000 acre-feet allotted both basins by the compact of 1922.

Of this amount Mexico was to receive 500,000 acre-feet annually until 1980 and 350,000 acre-feet annually afterward through the All-American Canal.

Within five years the United States was required to build a new regulating dam called Davis Dam, between Boulder Dam and Parker Dam, together with the works necessary to regulate the diversion of water to Mexico's diversion points at the international boundary.

Mexico also was to build a main diversion structure in the Colorado below the upper boundary line, connecting with the Mexican canal system near San Luis, Sonora. She was in addition to pay her share of the costs of Imperial Dam on the Pilot Knob section of the All-American Canal.

Lastly, the treaty was to be administered by the International Boundary Commission of 1889, whose name was to be changed to the International Boundary and Water Commission and whose members were to be accorded diplomatic status.

In submitting the treaty to the Senate for ratification, the secretary of state stated of it:

> The treaty is the product of long and patient negotiations on the part of both Governments. . . . It is an outstanding example of the settlement of international problems by mutual understanding and friendly negotiation. I cannot overemphasize its importance from the standpoint of international good-will, brought about not by the gift of any natural resource but simply by the application of those principles of comity and equity which should govern the determination of the equitable interests of two neighboring countries in the waters of international streams.

On January 22, 1945, hearings on the treaty began before the Senate Committee on Foreign Relations. Immediately a storm of protest arose.

Senator McCarran of Nevada viewed the treaty as "unjust, unfair and un-American."

Attorney General Kenney of California stated, "I do not think this treaty can stand the light of day," and advised the State Department to withdraw it from the Senate.

Phil D. Swing, former California congressman who had helped to draw up the Boulder Cañon Project Act, reported that the treaty violated its provisions.

The governor, senators and congressmen of California united to oppose it, and the state legislature appropriated $75,000 to block its ratification—the first time in our history that a state had appropriated funds to block the ratification of a treaty negotiated by the federal government.

The American Bar Association and the American Federation of Labor passed resolutions against the treaty.

And as a last public impetus the Metropolitan Water District of Southern California printed and distributed an attractive brochure declaiming against the treaty.

Why?

Boiled down, to pardon the simile, the cause of all this hue and cry was simply the 1,500,000 acre-feet of water allocated to Mexico.

And behind this, the grabbag tactics of California interests to control the major amount of water and power throughout the whole history of the development of the Colorado River.

As a requirement of the Boulder Cañon Project Act, California had been forced to agree by legislative action to limit her annual use of water to 4,400,000 acre-feet plus one-half of the surplus—about 1,000,000 acre-feet more. Hence out of the total 7,500,000 allocated to all the Lower Basin states, her maximum share was 5,400,000. And already she had made contracts for 5,362,000, divided among Imperial, Coachella and Palo Verde valleys, Los Angeles, San

Diego and the Metropolitan Water District by the Seven-Party Water Agreement of 1930.

Thus California was already getting nearly one-third of all the waters of the Colorado to which she does not contribute one drop; and she was not yet using more than a half of that third. Moreover, that California did not consider herself in danger had been shown by evidence presented by Senator Connally of Texas. The Imperial Irrigation District actually offered to supply Mexican needs out of its own facilities, but at a price of approximately $340,000 a year.

Previously opposed by Arizona, which in 1944 finally signed the Colorado River Compact, California was now renewing the struggle to obtain Mexico's share.

Her main argument that the pending treaty dangerously threatened a vital American water supply and gave top priority to Mexican landowners was patently absurd. Mexico's share is specifically limited to 1,500,000 acre-feet. This, as has been pointed out, is the approximate difference between the total average flow of 17,400,000 and the total 16,000,000 allocated to the United States. It comes out of the annual excess of over 7,000,000 now wasting unused into the gulf. It is less than half the amount that she originally requested; 300,000 less than she actually used in 1943.

Moreover, as is seldom pointed out, half or more of all the water allocated to Mexico by the treaty is "return flow" —water that has already been put to use within California and has seeped back into the river and canals on the way to Mexico. Therefore, the allocation to Mexico is but 3 or 4 per cent of the primary flow of the river.

The legal points are no less fallacious. California assumed that Mexico had no right to water conserved by Boulder Dam, since the Boulder Cañon Project Act provides that stored waters should be used "exclusively within the United States." Yet it also states that "nothing in this Act shall be construed as a denial or recognition of any rights, if any, in Mexico to the use of the waters of the Colorado System."

If California's efforts to block it were successful, Mexico of course could continue to increase her use of Colorado River water. She has every right to put this water to beneficial use. In the opinion of Sumner Welles:

> Under international law, the beneficial uses of water which a downstream country has built up cannot be cut down by new uses created at a later date by the upstream country. Should California interests later attempt to divert the amount of water required by such beneficial use within Mexico, and should Mexico then demand arbitration of this question, the United States is pledged by the Pan American Arbitration Treaty of 1929 to accede to such a demand. In that event, the weight of established law would be wholly on the side of Mexico.

The nation sympathizes with the phenomenal development of Southern California and her efforts to make Los Angeles one of the largest cities in the world. But California's unreasonable opposition to the treaty came at a most crucial time.

A movement was started in Mexico to deny California the migrant workers needed to harvest her wartime crops. All Latin America regarded the treaty as a test of the realities of our professed Good Neighbor Policy. The controversy cast a shadow over the Inter-American Conference on Problems of War and Peace then being held in Mexico City. Whether the treaty was ratified, then, was not only a measure of future relations between Mexico and the United States, but a threat to the hemispherical unity of the American republics achieved during the crucial war years.

It is pointedly significant, therefore, that in the spring of 1945 the United States ratified the treaty. Mexico's later ratification thus ends the long series of interstate, national and international controversies over the Colorado. It opens the way to a new era of development.

CHAPTER 5

Long View

A<small>ND</small> so for an instant we pause before the technological future.

Just four hundred years from the time the first white man sailed into its mouth, the wildest and most violently beautiful river in the world had been broken to the needs of man.

At Boulder the world's highest and largest dam had harnessed its destructive savagery and transformed it into electrical power. Below it at Parker the river had been plugged by the world's deepest dam and part of it drawn off to supply the country's fastest growing metropolitan area with water. Farther down at Imperial Dam it had been tapped by a great canal to irrigate the richest agricultural valley in America.

Combining governmental supervision with the initiative of private enterprise, the Colorado River Project had set the pace for the development of other great rivers in the United States by the Tennessee Valley Authority and others still to be created for the Columbia and the Missouri. It is showing, through the co-operation of Mexico and the United States, what eventually may be accomplished with other international streams like the mighty Amazon and the Congo.

But the Colorado River Project is still elastic. Only a start has been made. On the lower river Davis Dam and the Mexican diversion structure have yet to be built. Colorado-Phoenix diversion works to carry water to central Arizona are taking shape on blueprints. Costing $333,000,000 and generating 711,000 kilowatts, they will include the longest tunnels in the world.

On the upper river above Grand Cañon two more dams will be built to check the flow of silt into Lake Mead and to store water for diversion purposes. Each will rival in size giant Boulder Dam. Tentative plans for the first already have been made. Located in Bridge Cañon, it will cost $207,000,000 and have a generating capacity of 375,000 kilowatts of hydroelectric power.

So today we stand at the threshold of a new era whose extent no man can foresee. The last great wilderness of America is being transformed for modern man. Its vast arid deserts green and blossom. Its dark primeval forests gleam with lights. The roar of machines replaces that of the cougar in his lonely cañon, the whir and flap of fan belts those of wild hawks' wings. A total population still less than that of a single great metropolis slowly multiplies into the yet unborn, teaming multitudes of the future. And with all this, we steadily advance into a pattern of socialized living that replaces the uncontrolled individualism of the pioneer.

Man at last has conquered the land. But to what ultimate end no one can say. There is only a vague, inquiet feeling that in all his scheme of domination there is something he might have forgotten. It may well be that the river itself will have the last word, after all.

Capable of impounding the whole of the river for two years, Boulder Dam also retards its movement of silt. The water, settled and clear, is released; the silt remains. Within 50 years Lake Mead will be filled with water; in 300 years the whole vast reservoir behind the dam will be filled in solid with sand and silt. Another dam, probably in Bridge Cañon, must be built farther upriver. Then still another until all possible eleven sites will have been utilized and exhausted. Then again the Colorado will resume its way.

Which one of us dares assume that one transient race of men in its short span of a few hundred years can do more than retard for a geologic moment the river's immemorial and immeasurable task of transporting bodily the whole vast Colorado Pyramid into the sea?

We measure minutes. The river ignores millenniums.

In its time the whole continent has been submerged under the sea at least seven times. And each time that it has risen anew the river has resumed its task. Patiently it has carried back to their ocean floor the seashells beached on the summits of the loftiest peaks. Doggedly it has eaten away the mountains themselves.

This eternal palingenesis is the one vast blueprint of creation. In it the schemes of man seem insubstantial, indeed. Canals, dams, even whole cities vary only in comparison with those which have preceded them. None of them endure. The day may yet come when in a clogged jungle cañon or in the midst of a vast frozen tundra a new explorer from another civilization will stumble upon a strange ruin—a mighty wedge of hand-cast rock and its debris of twisted steel. Eventually its meaning will be deciphered. This, it will be told, was once called Boulder Dam. A relic of America's machine-age civilization. A colossal, crude contraption by which men dammed rivers to supply themselves with water, instead of condensing it from the clouds or other simple means.

Imagine a race of men so fertile of mind and cunning of hand but so blind of perception that it imagined the immutable laws of universal creation could be transgressed with impunity!

Why, a slight decrease in the infinitesimal amount of

carbon dioxide in the atmosphere would reduce world temperatures, cool the oceans, and bring on a glacial age. A slight increase would reverse the process and our deserts would be transformed into jungles.

For all our technological achievements, our very lives tremble upon the delicate scales of nature. We are as ultimately dependent upon the ancient verities of land and sky as were the prehistoric cliff dwellers. Man has not yet completed the full circle toward a realization that his own laws of life must conform in the long view with those greater laws to which he still and forever owes allegiance. In this awareness alone lies hidden the full story of the Colorado—its prehistoric past and its technological future.

PART FOUR
Grand Cañon

Grand Cañon

No WRITER of worth has ever seriously attempted to describe Grand Cañon; no artist has ever adequately portrayed it. None ever will. For while it is the most compelling single area on the earth's surface, it is not a landscape.

The regal ermine-cloaked Rockies; the somber moss-hung swamps and bayous of Florida and Louisiana; the romantic orange groves of California; the sweet clean meadows of the Ohio; the majestic bluffs along the Hudson; the poignantly beautiful prairies of Kansas; the dreamy plantations of the Deep South; the rugged grasslands of the Far west—all these and a hundred others offer true landscapes. Each has a distinctive tone, key, spirit and character which hold true and unique despite their infinite variations. They can be known, loved and partially expressed.

The Grand Cañon is beyond comprehension. No one could possibly love it. It is not distinguished by any one dominant quality. It is not unique in the individual sense. It is universal.

One cannot define humanity. One can only define the terms of humanity expressed by its many components: beauty, cruelty, tenderness, strength, awe, horror, serenity, sadness, joy. But to define life—the blended summation of all its infinite aspects—is impossible.

The Grand Cañon in nature is like the humanity of

man. It is the sum total of all the aspects of nature combined in one integrated whole. It is at once the smile and the frown upon the face of nature. In its heart is the savage, uncontrollable fury of all the inanimate universe, and at the same time the immeasurable serenity that succeeds it. It is creation.

Never static, never still, inconstant as the passing moment and yet endurable as time itself, it is the one great drama of evolutionary change perpetually recapitulated. Yet the cañon refutes even this geological reality. In its depths whole mountains contract and expand with the changing shadows. Clouds ebb in and out of the gorges like frothy tides. Peaks and buttes change shape and color constantly in the shifting light. None of this seems real. It is a realm of the fantastic unreal.

If I were forced to describe so sublime an immensity, I would define it with only one word: the ancient Sanskrit word for the nonexistent material world of the senses: *Maya*, or Illusion. It embodies all that man has ever achieved of the knowledge of reality: that all matter, as our own science now suspects, is but a manifestation of that primordial energy constituting the electron, whose ultimate source is mind, and hence illusory and insubstantial. Grand Cañon seems such a world. A world whose very mountains are but the shifting, dissolving, re-created thoughts of the One Omnipotent Mind. It is beyond sensory perception. It lies in the realm of metaphysics—the world of illusion, *Maya*.

No one is ever prepared for the cañon as one is for the gradually rising Rockies as he approaches across the plains. One simply crosses a flat plateau hirsute with cedar and great pines, and there at his feet it suddenly yawns.

The Rocky Mountains upside down; an immense intaglio instead of a cameo. A mountain chain, as it were, nearly 300 miles long, up to 18 miles wide, but a mile deep instead of a mile high.

Say this and there is no more to be said. See it in one look and there is nothing more to see.

A competent writer, justly noted in his generation for his appreciation of western America, once described the first

impact of the cañon by relating that he had seen strong men break down and weep. Sentimental bosh! I think more of the curt remark of that old Westerner who at his first sight is reported to have shrugged and said, "Now, by Jesus, I know where we can throw our old safety razor blades!"

Like all great things, the cañon takes time to appreciate. So be wary of your companion's instant rhapsody of applause. Be more wary of that sacred hush affected by others. Simply take your look, turn on your heel and leave. The cañon will be there if you ever return. And if you are drawn back, you will know then that it is a great experience not to be taken lightly, and not before you are ready for it.

El Tovar on the south rim, Fred Harvey's luxurious oasis at the terminus of the Santa Fe railroad spur and the highway from Williams, Arizona, has long been the stranger's starting point. From here convenient government roads lead to several other points of vantage: Hopi Point, 2 miles northwest; Yavapai Point, 1½ miles northeast; Grand View Point, 11 miles southeast; Desert View, 20 miles east; and Lipan Point, just off the road between these latter. At their tidy observation lookouts provided with telescopes, you can briefly encompass a faint idea of the main cañon's length of 217 miles, and its average width of 12 miles.

For the first time the absurdity of a letter to M. R. Tillotson, superintendent of the Grand Cañon National Park, will strike you as funny. He relates that a Hollywood movie director once wrote him, requesting aid in selecting a convenient, scenic spot where his cinematic hero could be filmed in the act of jumping a horse across the cañon. Actually the chasm is so long and wide that it has prevented the migration of animals to and from the forests on each side. Only on the Kaibab Plateau to the north are found the Kaibab white-tailed squirrels, the only species with ear tufts in the world. Parachutists have been dropped on certain buttes to search for prehistoric forms of animal life possibly isolated in the cañon. In this largest virgin forest in the United States roam queer dwarf burros which have strayed here, gone wild, inbred and become stunted. The north rim abounds in deer. A government trapper killed five hundred mountain lions in four years.

Winding down to the bottom are two easy horseback trails, Hermit Trail and Bright Angel. Descend one of these and the third dimension of the cañon begins to be apparent —its appalling depth. A sheer drop of one mile from the south rim, and 1,300 feet more from the north rim. Remembering that one mile in altitude is comparable to 800 miles of latitude, you can travel here the equivalent distance from central Mexico to northern Canada. It is a trail that drops from a snowstorm at the rim into semitropical weather at the river below. And one which leads through all the zones of plant life from the mesquite of the Lower Sonora Zone, through the Upper Sonora and Transition, to the quaking aspen of the Canadian Zone.

In length, breadth and depth the cañon grows. Its mere immensity takes hold. Yet these dimensions are but its frame. Like a drug, the more of it you take in, the more you want. Often riding through the forests along its rim I have come across lone wanderers held there a month, a year, a lifetime, by nothing more than its strange and indefinable quality of compelling fascination. Ostensibly they are vacationists and invalids, photographers and artists, mere sheepherders, old hunters and trappers—even a crackpot religious waiting for the world to come to an end. But it is the cañon that holds them. It is the most powerful mesmer I know.

What is there in it that exerts so universal an appeal? For one thing, it contains every shape known to man. Lofty peaks, whole mountains rise out of its depths. There are vast plateaus, flat-topped mesas, high buttes and monoliths. And all these are carved in the semblance of pyramids, temples, castles; of pinnacles, spires, fluted columns and towers; porticoes and abutments, bridges and arches, terraces, balconies, balustrades. They are solid and fragile, bare and covered with latticework and delicate carving. It is a stage that seems expressly built to contain in perpetuity appropriate sets for every dynasty, every religion, every legend and myth-drama that man has known—a vast universal depository, as it were, of mankind's structural and architectural heritage.

Cardenas and his men, the first we know to look into the cañon, saw in it shapes resembling the towers of their beloved Seville. Cardenas Butte is named for him; another for Coronado, who headed the first land expedition into the region; and Alarcon Terrace for the first ship captain to ascend the river.

Named for the pre-Columbian race of Mexico, which they conquered, is Aztec Amphitheater, and for the race that preceded it, Toltec Point. There is a point for ancient Centeotl, and one for Quetzal, which gave a name to the vanished, mysterious quetzal bird and the legendary, feathered serpent-god Quetzalcoatl.

For the Greeks there are temples named for Apollo, Castor and Pollux; for the Romans, the temples of Jupiter, Juno and Diana.

The Christian Bible is not forgotten. There is a temple here for Solomon more enduring than one made of Lebanon cedars, and another for Sheba that will last as long as the fable of her beauty.

Here, far from Egypt's land, is Cheops' Pyramid, a Tower of Ra, and the temples of Horus, Isis and Osiris.

There is a Persian temple for Zoroaster; Chinese temples for Mencius and Confucious.

The immortal Hindu philosophers, saviors and deities—perhaps the oldest known to man—have here as everywhere their proper shrines. There are temples for Buddha, Brahma and Devi; a Krishna shrine; and a temple for Manu, who throughout the destruction and rebirth of all continents, all worlds, watches over the progressive evolution of all life including that of man, its latest form.

Here as nowhere else are background and settings spacious and majestic enough for the great Germanic myth-drama. Across the titanic cañon to the Valhalla Plateau could race the winged steeds of the Valkyries carrying heroes killed in battle. There is Wotan's Throne; castles for Gunther and Freya; a lofty promontory named for Thor, with room to swing the mighty hammer whose blows echo back and forth from the cañon walls, and Siegfried's Pyre forever flaming in fiery rock.

So too are the English myth-dramas of the Arthurian legend and the Quest of the Holy Grail recapitulated in enduring stone. Here stands King Arthur's Castle with that of Guinevere. There is another for Sir Galahad; and one for the tragic maid Elaine to stand in, grieving at its casements for the peerless knight's return. There rises Lancelots' Point, there yawns Gawain Abyss. Holy Grail Temple still holds at dawn and sunset the light no man saw on its tragic quest. Here is a mighty stone named for the magic sword Excalibur—itself first drawn from stone. Still others are named for the magician Merlin, the traitor Modred.

Point after point emerges to mark all these, simply named for the people who have always known this as their traditional homeland: Apache Point for the mother tribe and others for its subtribes, the Jicarillo, Mescalero and Mimbres Apaches; still others for the Hopi, Navajo, Walapai, Pima, Yavapai, Papago, Cocopah and Comanche.

So they loom out of time and space, a named minimum out of the vast anonymous multitude. . . . What shape or form has man ever conceived of mind and built by hand that the cañon does not hold? Is that its secret which holds a watcher at its brim—to see foretold in it the yet unborn form of his wildest imagining, the shape of his secret longing?

Why, many a man has hardly noticed shapes in it at all. They are merely blobs of color. Color so rich and rampant that it floods the whole chasm; so powerful that it dissolves like acid all the shapes within it. Here, if you will, is a drama whose characters are colors: the royal purples, the angry reds, the mellow russets and monkish browns, soothing blues, shrieking yellows, tragic blacks and mystic whites, cool greens, pale lavenders and anemic grays.

A lifetime is too short to watch their infinite variations

in key and tone. They change with every season, every hour, and with every change in light and weather.

In the blinding glare of a summer's noon its tints are so muted that the cañon seems a delicate pastel. But watch it at sunset. The yellows slowly deepen to orange; the salmon pinks to reds; the greens and blue-grays to damson blue; the lilacs to purple. Sunrise reverses the process. The whole chasm lifts bodily, inch by inch, toward light. The paint pot tips and spills over. The colors run and seep down the walls, collecting in pools below.

If it is a picture, winter frames it best. Preferably after a heavy snowfall when the plateaus are solid white, and better yet when every twig and needle is still sheathed in ice. Deeply inset in such a frame the cañon has all the warmth and color of a child's stereopticon slide held up to the table lamp. Into it snow never descends. A summer rainstorm is more potent. Then mists and clouds are formed below. Like tiny puffs from father's pipe they spurt out of the warm cañons and swelling like balloons gradually float to surface.

But the cold, clear, cloudless days of October—that is its time. Its colors stand out flat and positive. They relate it, not to the universal, but to the earth in which it is set. Red Supai sandstone, the rich red rock with the Indian name, the bright red Indian earth that stains land and river alike and give both their name. Green Tonto shale, green as pine and sage, bright as turquoise, clear as the turquoise sky above. Red and green on limestone white. These are its distinctive colors as they are the colors of the old Hopi ceremonial sashes, the masks of the giant Zuñi Shalako, the Navajo blankets, the fine old blankets of Chimayo so faded with their lost and unduplicated colors.

In this, of course, I must own to a sentimental but helpless preference. This is my land and to it I belong with all it expresses and with all by which it is expressed. It is merely a matter of vibration. We are each keyed to that band of the spectrum which determines our own characteristic tone.

Little wonder then that the cañon is universal in appeal. It is the complete spectrum, and in its vast range there is no one who does not find his own harmonic key. Color is a mysterious thing. Within the written memory of man, as we know, there was a time when he could not distinguish between the blue of the sky and the green of the forest. Still today there are colors not all of us can perceive. But they are there in the cañon—a thousand gradations invisible and unnamed, yet each vibrating upon our consciousness. Is this the "music of the spheres" that fills us so with wonder, a celestial symphony of color that drives us to still another terminology to express the inexpressible feeling it evokes?

In an instant the whole thing is forgotten when suddenly after midnight a slash of lightning rips through the dark. One hears the bolt strike. It is as if it has cracked the hinge of a cañon wall. A cliff caves in. It tears down another, and it another, like the collapse of a pack of stacked cards. A tremendous and prolonged shattering, accompanied by a thunderous concatenation traveling down the whole cañon.

Before it is over you have thrown off your blankets and raced half naked to the rim. This is the end of the world as predicted by the crackpot religious. Bolt after bolt strikes into the gorge. In the hot dry air sheets of flame light up the crumbling buttes and peaks. A second later they have vanished, swallowed by a vacuous immensity of flame red and pitch black. It grows greater and greater to the echo of thunderclaps thrown back and forth from the remaining walls—an inferno bathed in fire, a chaotic underworld. This is the apocalypse, the most awful and most sublime sight you can experience. Before it you cling to a piñon, insensible to self, the shrieking wind and the lash of rain.

As suddenly, it is over. The last reverberation dies away. Overpowering silence breaks louder upon the eardrums. In this monstrous, unearthly calm the first light of day breaks over the clifftops. They are still standing. And in the clarity of rain-washed dawn you see a world reborn in the semblance of the old. But new, enthrallingly new!

Such a storm articulates that quality which subtly and powerfully impenetrates every other quality and every dimension of Grand Cañon. Time is its palpable fourth dimension. Yet its effect is indescribable. One can only stand mute and prosaically view its geologic record.

Grand Cañon is the world's largest and oldest book. It is over 15,000 feet thick and it contains the history of 2,000 million years. Though its pages are wrinkled, creased and worn, they are brilliantly colored and beautifully engraved. A few chapters are missing. But so clearly are the others written that their meaning is revealed without break in continuity.

Thumb down through its rock pages.

The forest-covered plateaus on each side terminate at the rims of the cañon in cliffs of light-gray limestone, almost white in sunlight, and filled with fossil shells, chert, agates and carnelians. This is Kaibab limestone, a layer 800 feet thick and named after the Kaibab Plateau on the north. Kaibab itself is a Paiute name meaning "Mountain Lying Down."

Under this is a stratum of Coconino sandstone darker gray in color and 300 feet thick. Its name comes from the Coconino Plateau to the south, whose rim is more than a thousand feet lower than that of the Kaibab.

Intruding below are massive bodies of red shales and ledges of red sandstone, 1,100 feet thick. This Supai Formation forms the wall of Cataract Cañon in which the Supais live, hence its name.

Below it stands a 500-foot wall of limestone called simply Redwall. It is almost pure calcium carbonate. Not only in texture does it differ from the shales and sandstone above, but in its shade of red. Originally blue-gray and still so when freshly broken, the wall has gradually been stained a dark red by wash and drippage from the overhead beds.

These four chapters, consisting of 2,700 feet of rock pages, form the upper part of Grand Cañon.

Below lies another part composed of two chapters called the Tonto group. The first of these is a layer of green Tonto shale 800 feet thick, and under it a thin 150-foot layer of basal Tonto limestone, much, much older than the Redwall above.

Below this opens the most complex and thickest formations in the whole book. It is aptly called the Grand Cañon Series, and it is 12,000 feet thick. The Unkar group comprises the first chapters. It is composed of brown Dox sandstone, dark Bass limestone, Hotauta basal conglomerate and Shinumo quartzite enlivened with bright red Hakatai shale. Such a mixed mass is well named in part at least; Shinumo was the name applied to the old Hopi confederacy. The Chuar group is somewhat similar. Its pages are torn, twisted and crumpled. For unlike the horizontal Tonto group, these formations are warped, folded and tilted, sloping to the north.

And now we see the last pages in Grand Cañon. The oldest rock system in the world, part of the original earth's crust. Great vertical layers of gneiss that formed before the earth had cooled, and huge blocks of granite forced into them in a molten state by heat and enormous pressure.

Such is Grand Cañon—15,650 feet and more of rock pages; pages of light gray, white, dark red, vivid green, blue-gray, bright red, brown; pages of coarse sandstones, fine-textured limestones, shales, rough conglomerates, quartzite, sturdy gneiss and granite. Never smooth, neatly pressed. But in horizontal layers, in vertical walls, in great folds. Warped, twisted, broken. Laid slantwise to encompass three miles of thickness within a vertical depth of one mile. And finally gouged out and eroded into a geological maze. What is the story it tells? Gradually we read back up the meaning of the pages.

For man time goes back no farther than the beginning of this geological record when the earth was still molten, but cooling to form its sturdy crust of gneiss and granite. This first Archeozoic era leaves no record of any primordial life for over a billion years, not even a single fossil. But gradually the earth cooled, atmosphere formed, water vapor condensed into rain, and the surface was eroded down into a plain.

The first chapter suddenly ends. This immense, immeasurable plain sank under the sea. Upon it were deposited the thick beds of sediment composing the Unkar and Chuar groups. First the limestone laid down upon the granite on the floor of the sea. Then the sandstone on its beaches. Finally the shale in its estuaries. For now something else was happening. The earth was rising again. Uplifted, it shook mightily, with vast distortioning shrugs, tilting and faulting the new layer on its crust.

So begins the next great era, the Proterozoic, which left in its seaweeds and crustaceans a record of the first life beginning on earth. But now again the persistent forces of destruction began—the solution of limestone by rain water, heat and cold, which expanded and contracted the rock, frost that cracked it, erosion by wind-blown sand. And again the surface was a rolling plain.

For a second time the earth sank under the sea. Again deposition of sedimentary beds took place—green muds of Tonto shale and limestone that buried even the small, hard-capped Unkar and Chuar islands.

But when it rose in Cambrian times, the third great

era had begun: the Paleozoic, the era of ancient life. Algae and seaweeds were abundant, trilobites, brachiopods and crustaceans.

There is a gap in the rock record here. Two whole chapters—the Ordovician and Silurian—are missing, during which shell-forming sea animals and reef-building corals developed. We know they exist; their formations have been found elsewhere throughout the earth. Of a third period there remain but a few isolated buttes; the Devonian, the "Age of Fishes," when amphibians and land plants began to form. Yet so clear is the pattern before and after this gap that there is little doubt that there were more submergences whose deposits here were washed away without trace.

The story resumes with another deep submergence during which the Redwall deposit was laid down. As the uplift slowly followed, the red Supai muds were deposited in the shallow water and then the Coconino sandstone on the beaches by strong currents. Each sand is distinctly formed and polished. This leaves little doubt that the uplift was exceedingly slow.

Still remains another submergence, deposit and uplift, showing the top layer of Kaibab limestone, which forms the present surface of the high northern plateau.

With this ends one of the great periods in geological history, the Carboniferous, in which all the oldest coal deposits throughout the world were accumulated and laid down under great pressure. It was the "Age of Amphibians," of sharks and sea monsters. Primitive flowering plants and the earliest cone-bearing trees began to form. Backboned animals came into being for the first time.

It ends too the third great geological era, the Paleozoic, 340 million years long. The end of all "ancient life" forms, as the Proterozoic was of all primordial life. And yet these last four formations—the Redwell, Supai, Coconino and Kaibab—took shape in its last period, the Carboniferous, and end the story of Grand Cañon.

What about the two following eras? The Mesozoic, era of medieval life, 140,000,000 years long, the "Age of Reptiles"—of the great land monsters, the dinosaurs, and the flying reptiles. The Cenozoic, era of modern life, the age of man, of modern animals and plants. Already it is 60,000,000 years in extent.

All around Grand Cañon remnants of its formations rise on the horizon—Vermilion Cliffs to the north, Cedar Mountain to the east, Red Butte to the south. All the rest has been swept away. Swept away by the great red river as it cut Grand Cañon. They were, in fact, the river's tools —every grain of sand, every frost-shattered boulder, every mountain peak. The teeth by which time has cut a visible gash through eternity.

This, then, is the story of Grand Cañon. The geological record of an earth at least 2,000,000,000 years old, nearly three-quarters of which had transpired before the first record of life, and all but 1,000,000 years before man came into existence. Over three miles thick it is, yet the whole of man's evolution is not represented by so much as the thickness of one sheet of paper.

We know so little, and we are so contemptuous of each other's knowledge. The oldest man here, the red man, has spoken in the science we call "myth." The white man speaks in the myth we call "science." Yet their identical story of these million missing years could be written on this paper in a few paragraphs.

In the Beginning, it would read, God created the world and the waters subsided and the earth rose. But in the eternal palingenesis, which proves an evolutionary scheme for living earths as well as for living plant and beast and man, this continent has had many such beginnings. Seven times, as we have seen, America has been submerged in the darkness of the deeps. And each time it has risen to stand in the light of a day millions of years long.

This day has hardly yet dawned.

In its first dim light only the summits of the highest peaks stood out. The little islands that are today the peaks and ridges of the Colorado Rockies, the oldest dry land on the American continent.

A moment more and the whole vast land rose shudderingly, streaming water from its flanks like an old buffalo from his wallow. It was a land whose size and shape and texture are hidden in the memory of an indigenous mythology not yet deciphered, and in a science not yet infallible to reasonable guess.

For suddenly it was changed. There began one of the greatest dramas America has ever known or will know until it disappears again from sight. The earth shook in a last convulsive tremor. Mountains split asunder. Peaks burst into flame. Whole upper halves of 30,000-foot volcanoes blew off, leaving only the stumps long afterward named peaks by men like Long and Pike.

And now the molten earth cooled, solidified. The last of the waters receded, hissing, leaving seashells and sharks' teeth imprisoned in rock. Whole forests lay level, petrified into stone. Mountains of debris washed down, packed solid; and still retaining their latent fire, became vast beds of coal. Wind and water gnawed at the rough-edged portrait of the land, changing it into the face we know now. Green life began, and that which fed upon it. But above all this stood out the ranges, shining with pristine smoothness, glitteringly white. The Shining Mountains, as they were first known and are still called by man.

And then the rivers began to form in their snowbanks and their glaciers. The hundreds of little rivers that drew into one. The great red river which cut its way down peak and mountain, through mesa and plateau, across the desert to the sea below. The river that has written these last few moments of earth-history.

Thus near dawn it stands, fresh, serene and virginal, yet old and worn and warped by the tortuous pain of rebirth—this vast river basin of a continent newly risen from the depths of time. The high, rugged Colorado Pyramid, which is at once the newest and the oldest dry earth of America.

And now breaks the first light of this, its new day. It strikes a few dark figures born of the earth itself who are watching a sail far off on the horizon. In a moment the ship enters the mouth of the river. A few strange figures step off—white. The two groups, white and dark, meet and intermingle for a moment on its shores. The shores of the river that expresses for both the common mystery of their lives and the earth they tread—the strange red river that flows and ever flows between them.

And still it is not yet dawn.

This is all we know of the Colorado's history. It is all that these paragraphs have been about.

Reference Appendix

To APPEND a complete list of references consulted would be both needless and misleading. It would fall far short of being a complete bibliography on the Colorado, one of the most written about and least known of rivers; and it would imply, like most imposing lists, an academic interest in its history which I have never had.

Controversies over whether the exact route of an early explorer lay to the right or left of a certain boulder have filled volumes. That his discovery of it was made on Monday or Friday is equally debatable. Names and localities, like Tusayan, have yet to be fully defined by undiscovered archives in Spain and Mexico. Many of the manuscripts we now have are contradictory. Translators worry over the wording of a phrase, scholars over its meaning. Religion is ever on the watch to guard the infallible sanctity of the church, particularly Catholicism and Mormonism. Patriotism, back-yard and national, distorts some versions; racial prejudice, still others. All this academic wrangling is dreary and meaningless. Earnest researchers but probe the lines in the face of the land, remaining myopically blind to its expressive moods and character.

Its people are in a worse fix. They have had to contend with the glib dramatizing of writers eager to curry quick public favor. Violent and romantic action is their dish. The devil takes ideas, cause and effect. These "Mistonamos," who, as the saying goes, stretch the truth to their own ends, are quick to make mountains out of molehills while blithely ignoring the greater peaks of stark reality. The Colorado River basin has always been a fertile field for such melodrama. But it has paid off its dues. The time has come for us to sink new shafts into the enduring drama behind all these, as year by year we perceive more clearly its real components. This book, then, is not an academic history of the Colorado. Nor has it been melodramatized to intrigue with local color all straying eyes. Its tales have long been told, and often. It is for us not to dress them up again, but to read their meaning.

Those references not mentioned in the text which have been of most help and were of most interest for various reasons include:

The publications of the Quivira Society, Los Angeles and Albuquerque, between 1931 and 1935, are particularly good background:

The Indian Uprising in Lower California 1734-1737—Father Sigismundo Taraval.
The Mercurio Volante of Don Carlos de Siguenza y Gongora.
An Account of the First Expedition of Don Diego de Vargas into New Mexico in 1692.
History of New Mexico—Gaspar Perez de Villagra. Alcala, 1610.
Diary of the Alarcon Expedition into Texas 1718-1719—Fray Francisco Celiz.
History of Texas 1673-1779 (2 volumes)—Fray Juan Augustin Morfi.

The American Guide Series, compiled and written by the Federal Writers' Project, for each of the seven Colorado River basin states, are the best regional references.

For geology: *Introduction to Historical Geology*—William J. Miller. I. Van Nostrand Co., New York, 1937.

Textbook of Geology. Part 1. Physical Geology—Louis V. Pirsson; Part 2. Historical Geology—Charles Schuchert & Carl O. Dunbar. John Wiley and Sons, Inc., New York 1929.

REFERENCE APPENDIX 383

Reconnaissance of Northwestern New Mexico and Northern Arizona—N. H. Darton. U.S. Geological Survey Bulletin #435.

Guidebook of the Western United States. U.S. Geological Bulletin #613.

Story of the Grand Cañon of Arizona—N. H. Darton. Fred Harvey, Kansas City, Mo., 1917.

The Titan of Chasms—C. A. Higgins, containing "The Scientific Explorer" by J. W. Powell. Chicago, 1902.

Ground Waters of the Indio Region—W. C. Mendenhall. U.S. Geological Survey, Water Supply Paper # 225.

Mojave Desert Region—U.S. Dept. of the Interior. Geological Survey, Water Supply Paper #378.

The best book on Lower California, wholly delightful, is:

Baja California al Dia—Aurelio de Vivanco. Privately printed, 1924.

Applauded in the text and worth inclusion here also:

Romance of the Colorado River—F. S. Dellenbaugh.

A Cañon Voyage—F. S. Dellenbaugh.

Other books include:

Brigham Young—M. R. Werner. (The standard work on the history of the Mormons). Harcourt, Brace & Co., New York, 1925.

The Colorado River—Lewis R. Freeman. Dodd, Mead & Co., New York, 1923.

The Salton Sea—George Kennan. The Macmillan Co., New York, 1917.

The California Deserts—Edmund C. Jaeger. Stanford University Press, 1938.

The Colorado Conquest—David O. Woodbury. Dodd, Mead & Co., New York, 1941.

And Still the Waters Run—Angie Debo (An excellent documented account of Indian injustices). Princeton University Press, 1940.

The White Heart of the Mojave—Edna Brush Perkins. Boni & Liveright, New York, 1922.

Rivers of America—Israel C. Russell. Putnam & Sons, New York.

The Early Far West—W. J. Ghent. Tudor Publishing Co., New York, 1936.

Brothers Under the Skin—Cary McWilliams (The best recent work on racial minorities). Little, Brown & Co., Boston, 1943.

Grand Cañon Country—M. R. Tillotson and Frank J. Taylor. Stanford University Press, 1929.

Pay Dirt—Glenn Chesney Quiett. D. Appleton-Century Co., New York, 1936.

Navajo Creation Myth—Hasteen Klah. Recorded by Mary C. Wheelwright. Museum of Navajo Ceremonial Art. Santa Fe, N. Mex., 1942.

Mesa, Cañon and Pueblo—Charles F. Lummis. The Century Co., New York, 1925.

Rio Grande—Harvey Ferguson. Alfred A. Knopf, New York, 1933.

The Colorado Delta—Godfrey Sykes. American Geographical Society. Special Publication No. 19. Published jointly in 1937 by The Carnegie Institution of Washington and the American Geographical Society of New York.

By far the best reference on the district, to which I am indebted to Miss Helen L. Williams for generously sending me her copy.

The Grand Cañon—Thomas F. Dawson. 65th Congress, 1st Session, Senate Doc. No. 42.

"An article giving the credit of first traversing the Grand Cañon of the Colorado to James White, a Colorado gold prospector, who it is claimed made the voyage two years previous to the expedition under the direction of Maj. J. W. Powell in 1869."

Not to be omitted are the following books of my own based on extensive research in each of their separate fields:

Midas of the Rockies (The biography of Winfield Scott Stratton and the history of the Colorado gold discoveries). Covici-Friede, New York, 1937.

The People of the Valley. (For historical and ethnological background of Spanish-Colonial settlements). Farrar & Rinehart, Inc., New York, 1941.

The Man Who Killed the Deer. (For Pueblo Indian life and ceremonialism). Farrar & Rinehart, Inc., New York, 1942.

For background on the beaver hat, the following books offer interesting and amusing sidelights:

Male and Female Costume—Believed to have been written by Beau Brummel. Reprinted in 1932 by Doubleday, Doran and Co., New York.
Habits and Men—Dr. Doran. Redfield Co., New York, 1855.
Accessories of Dress—K. M. Lester and B. V. Oerke. Manual Arts Press, Peoria, Ill., 1940.
Cyclopaedia of Costume—Planche. Costume Dept., Los Angeles Public Library.
Reports, Bulletins, Magazines, etc.
U.S. Dept. of State Bulletins: March 25, 1944; Jan. 21, 1945.
Reports of Colorado River Commission.
Report of Colorado River Board on Boulder Cañon Project—1928.
The Federal Government's Colorado River Project. Boulder Dam Association, 1927.
Boulder Dam—U.S. Dept. of the Interior, Bureau of Reclamation, 1937.
Subsidence of Colorado River Delta and its Prevention—J. O. Turle, 1928.
National Geographic Magazine: May, 1924; August, 1914; July, 1932.
Carta Semanal, Mexico D.F., Sept. 11, 1943:
"Serio Problema en la Baja California"—Julio Riquelme Inda.
Fortune Magazine—August, 1943.
From whose excellent account of the Six Companies most of my own information was freely drawn.
Life Magazine—Oct. 23, 1944.
"The Colorado"— a series of the best pictures I have seen of the river, which should be republished in pamphlet form.
To the Metropolitan Water District of Southern California I am indebted for the following publications:
History and First Annual Report, 1939.
Sixth Annual Report, 1944.
Colorado River Aqueduct, 1939.
Mexican Treaty, 1944.
Social Research—November, 1944.
Trail—September, 1917
"The Alleged Journey & The Real Journey of James White on the Colorado River in 1867"—Robert Brewster Stanton.

GLOSSARY *

agua caliente—warm water
agua fria—cold water
alcalde—mayor
algodon, pl. algodones—cotton
amagoquio—a chamois living in sierras flanking tidewater of Colorado delta
amatrada—a noise-making machine of toothed wood, used by Spanish Colonials of Sangre de Cristo mountains
aquacate—a semi-tropical fruit grown in southwest U. S. and Mexico
Arenoso Desert—"arenoso" is Sp. for "gravelly, gritty." Hence the name of this desert
arroyo—a watercourse, or a small gully or channel after water has evaporated
Aye de mi!—Woe is me!

bacanora—a clear, white liquor distilled from cactus juice
bajada—the road or path by which one descends
bancos—small tracts of land which were continually being separated from the U. S. or Mexico by change in the course of the Colorado River.
barranca—a deep gorge; hence, the overflow channel of the Colorado to prevent flooding of Imperial Valley
blanco, a—adj. white
braza—a fathom

calvario—Calvary
campesino—a countryman
cañon—a deep valley with high, steep sides
cantina—a canteen; a combined saloon and provision store
capita—small cloak
capitán—captain
cargadores—stevedores
casa blanca—white house
Casa del Desierto—House of the Desert
Casa Grande—Great House (the house of Montezuma)
cezontli—(Aztec) a bird like a woodpecker in mountains flanking tidewater of Colorado Delta. See also pajaro chollero
chaparral—a thicket of dwarf evergreen oaks; hence, any dense thicket of stiff, thorny shrubs or dwarf trees

chimbíca—Indian name for mountain lion
cholla—a form of bladed deerhorn cactus
chollo—a halfbreed
chusquatas—alternate name for tortillas used by Indians near Tarascan lake in Mexico
cinco—five
Como no?—Why not?
conquistadores—conquerors
copita—small glass or cup
criollo—a creole
Cruzados—name given by de Onate, ca. 1604, to Indians who wore little crosses
cuatro—four
cuatro y media—four and a half

dorado—a goldfinch

entradas, (pl.)—Sp. for "entrances, doors;" hence, the exploratory expeditions of Spanish Fathers which opened up unknown territory in the Southwest

fiesta—festival

galleta grass—1 inch desert grass in Southwest
Gran Quivira—A mythical town of fabulous wealth sought in 1541 by Coronado near the central part of present state of Kansas. The account originated from a story by an Indian captive.
guajes—gourds with handles

Hermanos Penitentes, or, Los Hermanos de la Luz—Brothers Penitent, or The Brothers of Light. A sect of Flagellants belonging to 3rd order of St. Francis. Brought to Mexico in 1598 by followers of de Onate, perished elsewhere, but survived in Sangre de Cristo mountains.
hogan—the Navajo house
hozhonji—Navajo word for "happiness," signifying harmony between man and the land
huaraches—Mexican sandals

ingeniero—engineer

kiva—underground chamber used for ceremonial purposes by cliff dwellers and pueblo Indians

* The Southwest is practically bi-lingual, so that many Spanish and Indian words and expressions are part of the everyday language. Unfortunately, there was no room to include in this glossary the English meanings of all Spanish and Indian place names, but some are defined in the text, and the meaning of others can often be guessed from the context.

GLOSSARY

Laguna Salada—Salty Lagoon
LLanos del Rio Colorado—"llanos" is Spanish word for "plain;" hence, the desolate coastal plain along eastern shore of Gulf of California
el lobo marino—sea wolf
loncherías—small, drab cafés in Mexicali

mesa—a flat-topped rocky hill having steeply sloping sides
Mesa Verde—Green Mesa
mesquite—a spiny, deep-rooted tree or shrub belonging to mimosa family, bearing bean-like pods rich in sugar, which grows in southwest U. S.
mestizo—a person of mixed Spanish and Indian blood
mezcal—a kind of pulque or maguey. All three are liquors, the fermented juice of the American agave plant. Pulque is national beverage of Mexican natives.
milpa—a maize field
morada—Sp. meaning "habitation." The thick-walled, windowless buildings used by Hermanos Penitentes for their rites.
El Morro—Castle Rock
muchacho—little boy
Los Muertos—The Dead

No entiendo—I don't understand

ocote torches—torches made from the ocote or torch pine
ocotillo—a thorny, scarlet-flowered candlewood, grows in southwest U. S. and Mexico

padre—father, or priest
paisano—peasant
pajarita—little bird
pajaro chollero—*see* cezontli
panocha—gluey brown pudding made from wheat sprouted, dried and then ground into flour
paseo—a walk, drive, ramble
la paz—peace
peon—a person bound to service in payment of a debt
permiso—permission
con su permiso, señor, (with your permission, sir) Excuse me, sir.
pez gallo—flying fish
picacho—mountains in Southwest having very sharp peaks
piedra—stone (Spanish)
pierre—stone (French); Les montagnes de pierres brillantes—The Mountains of Shining Stones, or The Shining Mountains
piñons—pine trees
Piños Altos—Tall Pines
pinto bean—a kind of mottled field bean of the same species as the kidney bean
pitero—man who plays the flute among Sp. Colonials of Sangre de Cristo mountains

pito—a reed flute, used by the Sp. Colonials of the Sangre de Cristo mountains
playas—vast, dry lake beds in desert
placita—a small, flagstone court
pobrecita—Spanish term of endearment, "Poor little thing!"
pocar robado—draw poker
pocar garanoma—stud poker
portal, pl. portales—porch, piazza, usually covered
presidios—military garrisons
pueblo—Spanish for "village"
punta—point (the geographical term)
la Purísma Concepción—the Immaculate Conception

Quien sabe?—Who knows?

ramada—Spanish for mass of branches, arbor
ramadas—the thatched homes of Cocopah Indians
ranchero—owner of a ranch
ratero—a thief, pickpocket
reata—a rope
rebozo—a muffler or shawl
rezador—Spanish, "one who prays often;" hence, the man who leads the chant in processions of the Hermanos Penitentes
rico y pobre—rich and poor
Rio Colorado—Spanish for "Red River"
rurales—police force in Imperial Valley

San, Santa—Saint
sangrador—Spanish for "bloodletter;" the man who cuts backs of Brothers Penitent in shape of a cross in their ceremony re-enacting the Passion
sangre de Cristo—blood of Christ
seis—six
sierras—mountains
sipapu—shallow hole in front of the firepit in a kiva, symbolizes the entrance into earthly regions from underworld of pre-birth.
soldado—soldier

taco—light lunch with draught of wine
tequila—Mexican liquor
tinieblas—"tiniebla" is Spanish for "darkness;" the last act of the Passion as re-enacted by Brothers Penitent—a moment of silence with everyone kneeling in darkness of their church, followed by tremendous burst of noise symbolizing the resurrection
tortillas—Mexican pancakes, made with Indian corn mashed and baked in a little pan
tule lands—overflowed land where a kind of bulrush (tule) grows

varsoviana—an old Spanish dance, still performed in remote mountain cañons by people of Spanish origin
vega—a flat plain
Virgen Santísima—most Sainted Virgin

Index

Aborigines, American, 74
Acoma, pueblo, 77, 158-59
Adams, Captain Samuel, explorer of the river, 234-36
Agua Caliente, Mexico, 311
Alamo Canal, 345
Alarcon, Hernando de, 139-40
Algodones, 345
Algodones Sand Dunes, 88
All-American canal, 321, 326, 353
Animal life, delta, 103
Animals and plants, desert, 92-93
Aqueduct, Colorado River, 347-50
Arenoso Desert, 88
Arizona, and river power, 326-28, 349
Ashley, William H., 170, 171, 173
Aunt Allie, story of, 217-21
Aztecs, 73

Basin. *See* Drainage area
Beatty, John C., 289-90, 298
Beaver, trapping of, 165-67
Bechtel. W. A., 332, 334
Beckwourth, James P., 172, 173, 180
Bent's Fort, 170
Berthoud Pass, 41
Birdseye, Claude H., voyage, 244
Black Cañon, 322-23, 336
Blankets, Navajo, 56
Border controversies, 314-16
Bores, tidal, 12, 116-19, 139-40, 178
  on voyage of the *Rio Colorado,* 116-19
  Topolobampo, wreck of, 104-5
Boulder Cañon, 322
Boulder Cañon Project Act, 324-28, 356, 357
Boulder Dam, 322, 325, 329-43
  construction companies, 333-35
  cost of project, 326
  financing and construction, 329-43
"Boulder Dam Association," 321-22
Boundary, United States-Mexico, 351
Bridge Cañon, 361
Bridger, Jim, 171, 172
  tall tales, 180
Brown, Frank M., 240-41
Brown's Hole, 173
Bruce, trader, 57-63
Bryce Cañon, II
Burr, Aaron, trial, 190
Burros, 29

Cabeza de Vaca, travels of, 133, 134
Calexico and Mexicali, 292
California, and Boulder Dam power, 327-28
  claim to river, 328, 356-357
  grabbag tactics, 356-57
  origin of name, 132
California Development Company, 290-95, 303
  control by Southern Pacific Company, 294, 298
Camels on the desert, 227-28
Canal, All-American, 321, 323, 326, 344-47, 353, 355
Canal builders (Hohokam), 72, 345
Cannibals of Tiburon, 125
Cañon. *See* names
Cañons unknown to the public, 343
Canyon Voyage, A, by Frederick S. Dellenbaugh, 239-40

Cardenas, Garcia Lopez de, discoverer of the Grand Cañon, 142-44, 371
Carson, Kit, 172, 175, 190, 250
Chaffey, George M., 290-91
Chandler, Harry, 321
"Cherokee," story of, 215-17
Chivington, Reverend J. M., 248-49
Cibola, Seven Cities of, 131, 133, 137, 140-41
Cities, large, 276-77
Cliff dwellers, 69-72
Cliff Palace, 73
Climax Mine, 41
Cochiti, pueblo, 78
Colonials, Spanish, 46-52
Color as geographical designation, 55
"Colorado," first use of name, 147
Colorado, admittance into Union, 206
Colorado Desert, 86-89
Colorado-Phoenix diversion works, 360
Colorado River Board, 326
Colorado River Compact, 321, 324, 325, 328, 352
Colorado River Controversies, 242
Colorado River Project, 317-18, 330, 352-53, 356
Compact, Colorado River, 320, 324, 325, 328, 352
Conquerors, the, 131-44
Continental Divide, 9
Coronado, Francisco Vasquez de, 139-44
Cortez, Hernando, 131-35
Cortez, Sea of, 123, 132
Cory, H. T., 294-95
Cotton, Mexicali Valley, 101
Creation of the world, Indian account, 160-161
Creede, Colorado, 208
Cripple Creek, Colorado, 210
"Crossing of the Fathers," 158
Crowe, Frank T., 331, 334
Cutch, India, earthquake, 118
"Cutting back" action of the river, 296

Damming and floods, 293
  See also Imperial Valley
Dams. *See* names
Dams, proposed, 322-23
Danby Dry Lake, 349
Dances, Indian, 35-36
Davis Dam, 355, 360
Dawes General Allotment Act, 252
Death in the desert, 94-95
Death Valley, 83-86
Dellenbaugh, Frederick S., *A Canyon Voyage,* 239-40
Delta, the, 4, 101-27, 353
  animal life, 103
Denver, Colorado Cañon and Pacific Railway, 240-41
"Deseret," Mormon State, 196
Desert, the, 81-100
  heat, 94-95
  highways, 98
  plants and animals, 92-93
Diaz, Melchior, 142
Discovery of river by Europeans, 140
Divide, Continental, 9
Dominican missions, 146
Drainage area, 3-8

Earp, Virgil, 218-19
Earp, Wyatt, 219-21

389

390 INDEX

"Earps, Fighting," 218–21
Earthquakes, 118–20
   in southern California, 85
   San Francisco, 295
El Dorado, 131
Enchanted Mesa (Katzimo), 77
Entradas. *See* Padres
Equilibrium, isostatic, 23–24, 119
Era of the Pacific, 272
Escalante, Father Francisco Silvestre Velez de, 152, 155–58
Estevan, story of, 136–38
Evolution, Indian account of, 162
Exploration, Navigation and Survey of the Colorado Rivers of the West, From the Standpoint of an Engineer, by Stanton, 242
Explorers, 189–92

Fascination of the mountains, 95–100
Fechin, Nicolai, 37
Fish in the Gulf of California, 121–22
Floods, and damming, 293
   danger in, 323
   late-spring, 295
   See also Imperial Valley
Flowers, desert, 92
   Rocky Mountain, 38–39
Franciscans, 146–47
Fred Harvey House, 281–83
Frémont, General John C., 190, 191
Frémont Pass, 41
Fur companies, 169–70
Furs, *See* Trappers, the

Gaines, Old Lady, 39
Galloway, Nathan, voyages, 242–43
Garces, Father Francisco, 152–55
General Rendezvous, trading center, 170, 173
Ghost towns, 25–32, 204, 206
Gila monster, 93
Gila River, 12, 148
Gold seekers, *See* Prospectors
Grand Canon, 63, 119, 365–78
   discovery by Cardenas, 143
   first crossing by white men, 158
   first man through, Major Powell, 237
   reported trip through, White's, 232–34
Grand Mesa (Mesa Grande), 67
Grand Valley Project, 318
Granite Reef dam, 318
Great American Desert, 82
Great Salt Lake, 195
Green River, 9, 238
Guadalupe, Russian village, 88–89
Guaymas, 126–27
Gulf of California (Sea of Cortez), 123, 132
Gunnison River, 9, 318
Gunsight Pass, 42

Half-Pint Petey, 279–81
Handcart Battalion, Mormon, 196
Handy, Lieutenant R. W. H., discoveries, 225–26
Hardy's Colorado, 226
Harriman, Edward H., 294, 295, 297, 298, 301
Harvey, Fred, House, 281–83
Hat, story of the, 164–65
Hearst, William Randolph, 321
Heat, desert, 94–95
Heber, Anthony K., 292–93

Hermanos Penitentes, Passion Play, 46–52
Highways, desert, 98
Hohokam (Canal Builders), 345
Hoosier Pass, 41
Hopis (Moquis), 147
   mesas, 154
Hydroelectric power, *See* Power

Imperial Canal, 323
Imperial Dam, 344, 359
Imperial Irrigation Company, 303
Imperial Irrigation District, 356
Imperial Land Company, 291
Imperial Valley, 119, 120, 287–316, 323, 346
   peril to, 324
   romance of, 351
Independence Mine, 210–11
Independence Pass, 41
Indians, Creation Myth, 160–61
   desert, 98
   evolution, Indian account of, 162
   in Latin America, 255–57
   leaders in war, 250
   massacre of, 248–49
   names of races, Indian, 258–59
   origin of, 74
   religion, 162–63
   risings against the padres, 146, 149, 154
   runners, 98
   taking of lands and decrease in population, 252–53
   wars with, 248–51
Inscription Rock, 148
International Boundary and Water Commission, 355
Intolerance, 270–71
Irrigation, 317–28
   largest area, 346
   see also Imperial Valley
Isostatic equilibrium, 23–24, 119
Ives, Lieutenant Joseph, 228–29

Jefferson, Territory of, 205
Jemez Plateau, 67, 75
Jenkins, telephone patrol man, 95–98
Jesuits, 146
Johnston expedition against the Mormons, 198–99
Journeys down the river, notable, 180
Junipero, Father, 153

Kahn, Felix, 332
Kaibab Plateau, 374
Kaiparowitz Plateau, 65
Kaiser, Henry J., 332, 334–35
Katzimo (Enchanted Mesa), 77
Kearny, Colonel Stephen W., 190–91
Kivas, sacred, 74, 161
Kolb, Ellsworth and Emery, voyages, 243–44
Kuehne, Father Eusebio (Eusebio Kino), 149–51

La Bomba, 117
Laguna Dam, 297, 323
Lake Creek Pass, 41
Lake Mead, 342–43, 360
Land of Standing Rocks, 55
Las Vegas, 336–38
La Veta Pass, 42
Leadville, Colorado, 29, 207
"League of the Southwest," 319
Lee, John D., leader in Mountain Meadows Massacre, 198, 240

INDEX 391

Length and volume of Colorado River, 13–14
Little Colorado River, II, 148, 153
Llanos del Rio Colorado, 89
Long, Stephen H., and Long's Peak, 190
Long view, 359–62
Los Angeles, and Boulder Dam power, 326–28
  earthquake, 85
  Metropolitan Water District, 327–28, 349, 350, 353, 356
Los Angeles Aqueduct, 353
Los Angeles *Times*, 321
Louisiana Purchase, 190, 191
Loveland Pass, 41
Lummis, Charles F., 191

MacDonald, Alan, 332, 334
MacDonald and Kahn, 332
Manly, William L., trip on the river, 230–32
Marble Cañon, 158
Marshall Pass, 41
Mason Charlie, 73
Maya (Illusion), 366
Mead, Commissioner of Reclamation, 342
Mead Lake, 342–43, 360
Mesas, 53–80
Metropolitan Water District. *See* Los Angeles
Mexicali, 306–10
Mexicali and Calexico, 292
Mexicali Valley, 353
  cotton in, 101
Mexico, negotiations with, over diversion of water, 293
  right to Colorado River water, 357–58
  ruins, 72
  Yaqui rising in Nogales, 312–13
Mexico-United States boundary, 351
Mexico-United States water treaty, 350–58
Middle river, exploration of, 236
Migration, westward, 186–89
Miller, Alfred J., artist, 174–75
Miners, 89–92
Mines, 18–23
Missions, 145–63
Mogollons ("Muggy Owens"), 63
Mojave Desert, 83
Monarch Pass, 41
Monument Cañon, 71
Moquis (Hopis), 147
Mormon Trail, 186, 191, 197
Mormons, 66–67, 192–204
  fanaticism, 200–1
  Handcart Battalion, 196
  Johnston expedition against, 198–99
  polygamy, 194, 196–97, 200
  obsolete, 201
Morrison-Knudsen Company, 331
Mountain Meadows Massacre, 198, 240
Mountains, 17–52
  fascination of, 95–100
"Muggy Owens" (Mogollons). 63

Names, early Spanish, for the Colorado, 140, 148, 151, 152, 157
National Irrigation Commission, 353
Navajo Indians, 71
  blankets, 56
  Creation Myth, 160–61
Navigability of the river, claim of Arizona, 326, 328, 349
Nevada and river power, 327–28
Nogales, Arizona, and Nogales, Sonora, 311–13
Nordenskjöld, Baron Gustaf, 73

Onate, Juan de, 147–49
Ore, 22–23
Oregon Trail, 186, 191
Osages, wealth, 263–64
Outcasts, the, 203–22
Outlaws, the, 213–22

Pacific, era of, 272
Pacific Bridge Company, 332
Padres, the, 145–163
Painted Desert, the, 55–56
Painted Forest, 63
Paradise, Arizona, 64
Paria River, II
Parker Dam, 327–28, 347–50, 359, 368
Passes, Alpine, 40
  Rocky Mountain, 40–45
Passion Play of the Penitentes, 48–52
Pattie, Sylvester and James Ohio, 176–80
Peaks, Rocky Mountain, 30
Penitentes, Passion Play, 46–52
Pike, General Zebulon M., 31, 189–90
Pike's Peak, 31
Pike's Peak Rush, 205–6
Pioneers, 181–84
Plants and animals, desert, 92–93
Plateaus, 53–79
Polygamy, *See* Mormons
Powell, Major John Wesley, voyages of, 237–40
Power, hydroelectric, 325
  use of, 326
Prehistoric ruins. *See* Pueblos
Presidios, 146
Prospectors, 89–92, 203–13
Prospectors, fate of, 211
Pueblos, 69–72
Pyramid. Colorado, 4–8
Pyramid City, 278–84

Railroads, 282–83
Rainfall in the river basin, 7
Randolph, Espes, 294
Reclamation, Colorado River project, 317–28
Religion, Indian, 162–63
Rendezvous, General. *See* General Rendezvous
Rigdon, Sidney, 193
Rio Colorado (steamship), voyage on the lower river, 105–27
Rio Grande, water from, 351–52
River system. *See* Tributaries
Roads, desert, 98
Robbers, stagecoach, 63–64
Rockwood, Charles Robinson, 289–95 298, 303
Rockwood Heading, 323, 345
Rocky Mountains, 31–52
  peaks, 30
  ranges, 42
  snowstorm in May, 43–45
Roosevelt storage dam, 318
Royal Gorge of the Arkansas, 41
Ruins, prehistoric. *See* Pueblos
Runners, Indian, 98
"Running the river." 227–44
Russian village, Guadalupe, 88–89
Ruxton, Frederick, 180

Salt River, 12
Salt River Project, 318
Salton Sink and Salton Sea, 87, 143, 293, 294, 295, 299

Sand Creek massacre, 248–49
San Francisco, earthquake and fire, 295
San Gorgonio Pass, 348
San Juan River, 10
Santa Fe Trail, 186, 191
Sea of Cortez (Gulf of California), 125, 132
Sedimentation. See Silt
Settlers, the, 185–202
Seven-Party Water Agreement of 1930, 357
Shallow Water, trading post, 57–63
Shea, J. F., Company, 331
"Shining Mountains" (Rockies), 31
Silt, and effect of, 14, 118, 120–21, 360
Six Companies, 333, 335, 337, 339, 342
Smith, Joseph, founder of Mormonism, 192–96
So Boulder Was Built, 342
Solid content of river, 13–14. *See also* Silt
Southern California Edison Company, 326
Southern Pacific Railway, and the Imperial Valley, 292, 293–300
  control of California Development Company, 294, 298,
  repayment by Federal government, 300
Spanish-Colonials, 46–52
Stagecoach robbers, 63–64
Stanton, Robert Brewster, explorer of the river, 233, 240–42
Steamboating on the river, 227–30
Stieglitz, Alfred, 33
Stratton, W. S., 210–11
Strawberry Valley Project, 318
Street, Julian, 19–20
Sunset Dance, 35–36
Supai Indians, 374
Swing, Phil D., 356
Swing-Johnson Act, 324–25, 346

Tabor, H. A. W., 207
Tabor, Mrs. H. A. W., 27–29
Taos, New Mexico, trading center, 170
Telephone, desert, 95
Tesla, Nikola, 325
Tiburon, home of cannibals, 125
Tijuana, Mexico, 310–11
Tillotson. M. R., 368
Tombstone Travesty, 216–20
Tonichi, village of, 150
Topolobampo (steamship), wreck in the bore, 104–5
Towns, ghost, 25–32
  Imperial Valley, 292
  Pyramid City, 283–84

Trading centers for trappers, 170
Trails to the west, 186
  See also names
Trappers, the, 164–84
Travelers on the river, 224–44
Treaties for uses of rivers, 353–54
Treaty, United States-Mexico, for use of water, 350–58
  opposition to, 355–56
Tributaries, 8–13
Trueworthy, Captain Thomas, 234–36
Tuberculosis, 98–100
Turle, J. O., studies of the delta, 119, 120

Ulloa, Francisco de, travels, 133–34
Uncompahgre Project, 318
United States-Mexico boundary, 351
United States-Mexico water treaty, 350–58
Utah, and river power, 326
  See also Mormons, the
Utah Construction Company, 331
Ute Pass, 41–42

Vérendrye, Pierre Gualtier de Varennes de, 31
View, long, 359–62
Virgin River, 12
Volume and length of the Colorado, 13–14
Voyage, lower river, 105–27
Voyagers on the river, 224–44
Vrain Girls, the, 57–63

Water treaty, United States-Mexico, 350–58
Wattis, W. H., and E. O., 330–31, 334
Wattis-Morrison group, 331–33
Welles, Sumner, on river water rights, 357
Westerner, professional, 273–74
Wetherill, Dick, 73
White, James, trip on the river, 232–34
Wild Flower Special (train), 38–39
Williams, Bill, 183–84
Woman who rode away, the, 57–63
Wozencraft. Dr,. Oliver Meredith, 288–89
Wright, Harold Bell, *The Winning of Barbara Worth*, 302, 305

Yaquis, 312
Young, Brigham, 194, 198
  polygamy, 200
Yuma Desert, 88
Yuma Project, 318